MASS MEDIA AND FARM WOMEN

MASS MEDIA AND FARM WOMEN

(Foreword by C. Prasad)

REKHA BHAGAT
P N MATHUR

INTELLECTUAL PUBLISHING HOUSE
23, Daryaganj, Ansari Road,
New Delhi - 110 002

Sole Distributors

Intellectual Book Corner,
23, Daryaganj, Ansari Road,
New Delhi - 1100 02

Copyright: Rekha Bhagat & P N Mathur

ISBN No. 81-7076-019-4

First Edition: February, 1989

Price: Rs.   75.00
         US $ 15.00

All rights reserved.   No part of this publication may be reproduced in any form or any means without the prior permission of the publisher.

Printed in India
Published by D.R. Chopra
INTELLECTUAL PUBLISHING HOUSE
and Printed at Chaman Offset Printers,
1626, Suiwalan, New Delhi- 1100 02

Composed at
RUCHIKAM PHOTO SERVICES
229, Hauz Rani, Malviya Nagar,
New Delhi - 1100 17

Printed at **Chaman Offset Printers**, 1626, Suiwalan,

CONTENTS

Foreword			vii
Preface			ix
Chapter	I	Introduction	1
	II	Farm Women - A Profile	5
	III	Past Research Work	16
	IV	Studies on Newspapers and Magazines	29
	V	Studies on Films and Multi Media Approach	33
	VI	Research Strategy	40
	VII	Radio the Most Powerful Medium	50
	VIII	Television - Medium of the Future	57
	IX	Print Media - a Limited Reach	65
	X	Films Popular but Controversial Medium	70
	XI	What do Farm Women Think about Social Issues	77
Appendix			85
Bibliography			92
Index			101

FOREWORD

Of late, women in rural areas in general and farm women in particular has become a global subject for discussion. And why not? They have been neglected, ignored, and discriminated on all fronts by the male-dominated society everywhere but more so in developing countries.

The constitution of India provides for equal rights and privileges for men and women and makes special provisions for women to help them to improve their status in society. A number of social enactments have been introduced for removing various constraints which hindered their progress. No doubt that the constitution contemplates a social revolution brought out through the use of law as an instrument of directed social change. But in spite of these, there has been staggering progress in ameliorating the working conditions and socio-economic status of women especially in the rural areas. There has been even lack of awareness among the people about various legislations and programmes being implemented for the benefit of women.

The census of India estimated an all-India economic participation rate of 21% for women and 53% for men in 1981. Nearly 63% of all economically active men were engaged in agriculture compared with 78% of women. Almost 50% of the rural female workers were classified as agricultural labourers and 37% as cultivators. The respective proportions of male rural workers have been reversed with 55% reported as cultivators and only 24% as agricultural labourers.

Since 1970s, a serious concern for women's emancipation in all spheres has been expressed. In India such efforts have also been made by different Ministries/Departments and Non-Government Organizations, but in a vast country like India, the impact of such efforts have hardly been enough. In case of agriculture where involvement of women is far more prominent, the situation is no better. In fact, the agricultural development system has not yet taken farm women population seriously in the main stream of its development process. Therefore, it is high-time that this issue is discussed and deliberated at the national and regional levels in order to sensitize people about its potentiality, prospects and obstacles, and charting the course of action devoted to involving women as active partners in research and development with equal rights and privileges.

The Research system in agriculture has also not made deliberate efforts in tackling specific problems which confronted farm women. One argument has been that the research is gender neutral and the research results can be utilized by the farm men and women alike. Perhaps, this may be possible to some extent, but a well guided and directed research on operations/areas in which farm women are involved can fetch added dividends. There is also a need for culling and identify research results in all disciplines which are readily available for the extension system for farm women.

Indian farming community represent a complex social system including its heterogeneous nature, economic and social inequalities, caste and class differentiation, traditional social structure and systems, regional imbalances in development, under-developed village institutions, very low level of education, and a relatively fixed and expected roles of women. To cut across these barriers, a well coordinated, directed and concerted efforts by all agencies responsible for and interested in rapid development of rural women have to be made at all levels of administrative and development hierarchies.

Women are great communicators. This unique ability of women need to be utilized in dissemination of appropriate technologies in agriculture as well as in other sectors of development. Besides, all media of communication - Radio, Television, films and Print, must play a decisive role in keeping women abreast with the latest. The extension system needs to play added attention on rural/farm women in its three dimensional approach - farm advisory services, communication support, and training programmes. I compliment the authors of this book who, based on an empirical study, have thrown sufficient light on Mass Media and Farm Women. I am sure, readers will find it informative, useful and productive.

C.PRASAD
Deputy Director General
(Agricultural Extension)
I.C.A.R.

PREFACE

Rapid expansion of mass media has introduced a new element in the development strategy being followed in rural India. Media reach and the subsequent impact on the rural audience has become a subject of debate. Often experts have talked about urban bias in the media and have cast doubts about their impact on the rural population. The so-called socio-psychological impact is although often mentioned but rarely studied. This is more true when farm women are considered as an important segment of rural society. This book is a realization of the fact that farm women are a category in themselves and the media relationship of this category need to be probed into. The first author selected this area of research for her Ph.D. work at the Indian Agricultural Research Institute, New Delhi. Since the findings of the study kindled interest in many social scientists and media experts, it was decided to publish it in a suitable form. But keeping in view the information needs of the potential readers of this book, several additional informations have been addeed. For example, the chapter on profile of rural women has been expanded to provide demographic and other data which are normally not available at one place. Information on studies conducted on different media is a qunique collection of research studies conducted at important centres of media education, media production and media training.

Chapters on different media of communication have been further enriched with the latest available information. Efforts have been made to provide suitable background information on each media of communication so that students interested in studying these media in various contexts could get an idea of their background, reach, programme production, impact etc. The chapters on impact of media on opinions on socio-economic and other issues is perhaps one of the pioneering efforts to reflect the mood of the farm women vis-a-vis their environment. In fact, study of changing life-style of farm women itself is an interesting phenomena of which a glimpse has been presented here.

Two important features of this book are (i) the collection of references which covers a wide range of research studies, text books, official reports, records and several other documents, and (ii) a scientific measurement of life-style of farm women. This measurement developed by the authors can be effectively used by researchers and others who are interested in probing into the socio-psychological aspects of life of rural women.

Admittedly, this line of research pursued by the authors is an initial attempt at unfolding media -- farm women relationship. The women dimension need to be studied from various angles. Therefore, the researchers should accord priority to this area and come out with additional facts useful to the media and extension experts.

The authors are grateful to many friends in Delhi villages who have helped in conducting this study. The contributions of Dr. C. Prasad, Deputy Director-General (Agri. Extension), ICAR and our colleagues in the Division of Agricultural Extension, IARI are duly acknowledged. Our thanks to Maj. H.R. Chopra for promptly agreeing to publish this book in time, Mr. S. Moitra, Chief Librarian, ICAR for helping us in finalising the references and Mr. Surrinder Kumar for typing the manuscript.

Rekha Bhagat.
P.N. Mathur

Chapter I

Introduction

Developments in the field of Communication have greatly facilitated the spread of messages of change. The communication network has reached even the remote rural areas of the country. In India, where 75 per cent of the people live in over 5 lakh villages, speak several languages and maintain diverse cultural identities, mass-media of communication assume very important responsibilities. Realising that mass media are important precursors to change, improving media-reach in rural areas has become a guiding factor in rural development strategy.

The cause and effect relationship between man and media does exist but establishing this relationship is a difficult proposition because of methodological reasons. Nevertheless, increased use of mass media is considered as the first step towards modernisation (Lerner, 1958 and Rogers, 1969). Researches have shown that mass media can have tremendous persuasive impact on people and can be employed as dependable devices in furtherance of modernisation among people. In the present context of development in India, it is worthwhile to probe into the process through which change is creeping in different segments of rural society.

Farm Women

Although impact of mass media on rural population has been a subject of interest to many development strategists, only few efforts have been made in studying this relationship. Studies on the impact of media on rural women have been almost negligible. This probably has been an unintentional lapse since farm women are thought to be untouched by mass media. It was perhaps right to believe so in the past when the tentacles of media had not touched even the rural male folks. But with each passing day the media scenario is changing fast and more people, including farm women, are coming within the folds of the media messages.

It is natural for experts to be interested in the studies on farm women since they form the bulk of our population; in fact, they are half of our society, half of our family and half of our labour force. Keeping them away from the mainstream of development will retard the process of social change. Practitioners are therefore greatly concerned about involving farm women in the process of rural development. Obviously, rural woman is an important subject of several socio-economic studies and discussions in seminars, workshops etc. all aiming at making them equal partners in developmental efforts.

Traditional lifestyle of farm women of the past is slowly undergoing a change. Earlier, their life was traditional with very little freedom in their lives. They were home-bound with poor status. They hardly had rational views and opinions on important social issues because of lack of exposure to the outside world. However, in the recent past, many forces have started operating in the villages which affected their lifestyle. Some of the forces are: availability of school education, rural development programmes, transport facilities etc. resulting in more urban contact, better chances of getting jobs in urban areas and higher exposure to mass media. Farm women could not remain untouched by such influx of new agents of change. Their lifestyle started undergoing a change and they became aware of important social happenings and some of them even forming definite opinions on them. Their consumption pattern, dress and behaviour at home and in society underwent a change.

All that is happening around them which changes their environment and lifestyle needs to be studied in detail. Obviously, many things are happening in rural society and the lifestyle of farm women is undergoing a rapid change. What kind of changes are taking place and how, is a subject of study.

Very few studies are available which highlight relationship between media and women. Some studies in India were conducted to explain the media as source of information, media-use behaviour and programme preferences. At micro level, researchers in the field of Agricultural Extension have studied gain in and retention of knowledge and attitude change. Influence of media on life style and that too of farm women was not a favourite subject of study. Such influences on individuals irrespective of sex were investigated by Learner (1963), Rogers (1969), Mcnelly (1970) etc. They all tried to associate personal characteristics of the people with modernization. Committee on Status of Women in India (1974) was a major empirical survey conducted for highlighting the position of women in society.

However, this survey did not aim at establishing any cause and effect relationship.

The empirical data which forms the basis of this book has been generated through a study which was conducted by the authors in three villages of Delhi territory. The main aim was to establish possible relationship between mass media and changing life style of farm women. Although Delhi villages are not representative in character, the data generated have shown a glimpse of the future situation which will develop in other villages as well. The objectives of the study were as follows:-

i) To study the extent of mass media exposure and life style of farm women and related factors,

ii) To study the opinions of farm women about the impact of each media on the life style of farm women,

iii) To study the changing opinion pattern of farm women towards selected social issues related to day-to-day living.

The major hypothesis was that there is an impact of higher mass media exposure on the lifestyle of farm women. The components of lifestyle included quality of life, work and activities, media use, community identification, social attitude and family situation. The mass media included in the study were radio, television, print media including newspapers and magazines and films.

The quality of life indicates habits and tastes in what women possess, quality of daily routine, religion in life and concern for children's education. Work and activities include involvement in community affairs, liking for trying out and learning new things in life, taking pride in appearance and self, enjoying more leisure etc. Community identification, the extent farm women prefer a community where people interact with others and have a 'us' feeling as member of the community. Social attitudes included attitude towards household work, marriage style, marriage related issues, children's education etc. A typical modern farm woman was hypothesised as a woman with a reasonable high level of mass media exposure having positive attitude towards important social issues of day-to-day life, having progressive outlook with high community awareness and interaction.

The findings as presented in this book form the basis of different aspects of farm women's mass media exposure, lifestyle and opinions on social issues. The relevance of data lies in the fact that these have been collected after rigorous scientific exercise and are authentic. The conclusions drawn indicate towards a definite trend in changing lifestyle and media utilisation as emerging in different rural settings in India.

Chapter II

Farm Women - a Profile

A national Perspective

India being the second largest country in the world in terms of population, also has a large women population. The estimated female population of 331 millions (1981), which is about 46 percent of the total population, is more than the total population of many countries. To understand the relationship between women and development, it is essential to have an analytical look on the demographic aspects of women, especially the rural women.

Sex-Ratio

The birth of a female baby in the family is not generally a matter of celebration in India. That is why there has been a constant decline in the sex ratio. While in 1901, there were 972 females for every 1000 males, in 1981, the ratio declined to 933 females to 1000 males. While there were 331 million females, the number of males was nearly 354 million. Table 1 shows the growth of female population in India from 1901 to 1981. It may be noted, barring the age group of 20-29 years, there are more males than females in all the age groups in India.

The declining sex ratio may be mainly due to female mortality which is higher than males even in the age group of 15-29 years. This is true of both rural as well as urban areas. Most of the female deaths in this age group are due to pregnancy or child birth. Table-2 provides detailed information on average expectation of life at birth (in years).

Table 1
Growth of Female Population in India 1901-1981 (in millions)

Year	Total population	Female population	Females as % of total population	Females Per 1000 males
1901	203	117	49.2	972
1911	252	124	49.2	964
1921	251	123	49.0	955
1931	279	136	48.8	950
1941	319	155	48.6	945
1951	361	175	48.5	946
1961	439	213	48.5	941
1971	548	264	48.2	930
1981	685	331	48.3	933

Source: Census of India, 1981.

Table 2
Average expectation of life at birth (in years)

Year	Male	Female
1901/11	22.6	23.3
1911/21	19.4	20.9
1921/31	26.9	26.6
1931/41	32.1	31.4
1941/51	32.4	31.7
1951/61	41.9	40.6
1961/71	46.4	44.7
1971/76	50.1	48.8
1976/81	52.6	51.6
1981/86	55.1	54.3
1986/91	57.6	57.1

Source: Health for all: An alternative strategy (ICSSR-ICMR); and Economic Situation and Prospects of India, World Bank, 1982.

Work Force and Occupation

Females represent nearly 46 per cent of the Indian population and nearly 80 per cent of the economically active women are engaged in agriculture compared to 63.3 per cent men. Out of total female workers, 46.2 per cent are classified as agriculture labourers and 33.3 per cent as cultivators.

There seems to be a traditional division between men and women in rural society. Farm women have the dual responsibility of working on the farm and in the home. Yet, their work participation rate is much less as compared to males. Table-3 gives comparative analysis of work participation rates for males and females in rural and urban areas.

Table 3

Work participation rates for male and female workers 1971 and 1981

	Males 1971	Males 1981	Females 1971	Females 1981
All ages				
Total	52.61	52.66	12.06	13.99
Rural	53.62	62.62	13.36	16.00
Urban	48.80	48.54	6.65	7.28

Source: Census of India, 1981.

The details of the occupational pattern of female workers of 1971 and 1981 can be seen in Table-4.

As per 1981 census, 80 per cent of female workers are engaged as labourers or cultivators. A third of all women workers are cultivators and almost half are agricultural labourers. According to a FAO report, 50 per cent of farm production is attributed to farm women. Sadly, within the farm sector, women are alloted work involving greater drudgery. Obviously, increasing their participation rate is not enough to raise women's status. They need to be educated and given more meaningful tasks in agriculture sector.

Table 4

Occupational pattern of female workers
1971 and 1981

	1971 Males	1971 Females	1981 Males	1981 Females
A. *Agricultural Workers*	69.68	82.61	65.60	81.23
Cultivators	45.90	29.84	43.70	33.20
Agricultural labourers	21.54	50.86	19.56	46.18
Livestock, forestry, fishing, etc.	2.24	1.91	2.39	1.85
B. *Non-Agricultural workers*	30.32	17.39	34.40	18.77
Mining and quarrying, manufacturing, processing, servicing and repairs.				
a) Household, Industry	3.42	4.24	3.18	4.59
b) Other than Household industry	6.70	2.77	8.92	3.55
Construction	1.36	0.65	1.81	0.80
Transport, storage and communication	2.86	0.47	3.32	0.38
Other services	9.07	7.08	9.22	7.05
Total	100.00	100.00	100.00	100.00

Source: Census of India, 1981.

Education

Education of which Indian women are deprived, is a prerequisite for faster development. As indicated by 1981 census, low literacy rate prevails in India both in case of males and females. It was 25.68 per cent for females as compared to 54.84 per cent for males. The rural female literacy rate in India is 17.57 per cent (See Table 5).

Table 5
Adult literacy rate in India

	1971			1981		
	Persons	Males	Females	Persons	Males	Females
Total:	30.04	47.69	19.32	40.76	54.84	25.68
Rural	26.98	40.51	12.88	32.70	47.27	17.57

Source: Census of India, 1981.

Besides variation in literacy rates of urban and rural females, inter-state differences also exist. Of every 100 adult women, as many as 64.48 are literate in Kerala, corresponding figures are much lower in other states (See Table 6).

Table 6
States and Union territories in order of female literacy, 1981

States	Per cent female literates to total female population
	Female literacy 50 per cent and above.
Kerala	64.48
Chandigarh	59.30
Mizoram	52.57
Delhi	52.56

Contd...

Table 6 (Contd....)

States	Per cent female literates to total female population
	Female literacy 25-50 per cent
Goa, Daman and Diu	46.78
Pondicherry	44.30
Lakshadweep	44.21
Andaman and Nicobar Islands	41.85
Maharashtra	35.08
Punjab	34.14
Tamil Nadu	34.12
Nagaland	33.72
Gujarat	32.31
Tripura	31.60
Himachal Pradesh	31.39
Manipur	30.69
West Bengal	29.28
Karnataka	27.83
	Female literacy less than 25 per cent
Haryana	22.23
Sikkim	22.07
Orissa	21.11
Andhra Pradesh	20.52
Dadra and Nagar Haveli	16.75
Madhya Pradesh	15.54
Uttar Pradesh	14.42
Bihar	13.58
Rajasthan	11.32
Arunachal Pradesh	11.02

Source: Census of India, 1981.

It must be noted that female literacy has made great strides in the twentieth century. From 0.6 per cent in 1901, it came up to 25 per cent in 1981 (See Table 7).

Table 7

Trends in female literacy in India (per cent)

Year	Males	Females
1901	9.80	0.62
1911	10.6	1.1
1921	12.2	1.8
1931	15.6	2.9
1941	24.9	7.3
1951	25.0	7.9
1961	34.4	13.0
1971	39.5	18.7
1981	46.9	24.8

Note (a) The literacy rate is percentage of literates to total population, exclusive of population in the age group 0-4 years. The rates up to 1941 are for undivided India.

(b) Excludes Assam.

Source: Census of India, 1981.

Male literacy has also progressed but at a much lesser rate. As a result the percentage of male literates which was more than 16 times than that of female literates in 1901, was less than twice in 1981. The gap is still quite wide.

The relationship between education and development at individual's level has been repeatedly established e.g. correlation between female literacy and infant mortality is equally clear (See Table-8).

Table 8
Infant mortality level by education of women in rural and urban areas, 1978

Educational level of women	Infant mortality rate Rural	Urban
Illiterate	175	88
Literate but below primary	101	57
Primary and above	71	47
Total literate	90	50

Source: Survey of Infant and Child Mortality 1979, Office of the Registrar General.

It can be seen that infant mortality happened to be highest in the case of illiterate mothers both in urban and rural areas but much higher in latter. As education level improved, the mortality rate declined. This presents a strong case for improvement in the literacy and education levels of women especially in rural areas.

Women recorded notable progress in higher education as is revealed by Table 9.

Table 9

Educational levels of males and females in India, 1981.

	Males Lakh	%	Females Lakh	%	Total Lakh	%	Population (excluding 0-4 years)
Literates with no educational level	1,139	90.6	531	66.6	1,670	69.3	28.7
Primary	438	30.3	271	34.0	760	31.5	13.1
Middle	290	18.0	134	16.8	424	17.6	7.3
Matriculation or Higher Secondary	279	17.3	100	12.5	379	15.7	6.5
Non technical diploma or certificate not equal to degree	2	0.1	1	0.1	3	0.1	0.1
Technical diploma or certificate not equal to degree	8	0.5	2	0.3	10	0.4	0.2
Graduate and above	71	4.4	23	2.9	24	3.9	1.6
Literates without educational level	474	29.4	226	33.4	740	30.7	72.7
Total	1613	100.0	797	100.0	2410	100.0	41.4

Source: Census of India, 1981.

Family Planning:

Government of India is popularising the concept of small family norm in a big way. According to data from the Department of Family Welfare, Govt. of India, between 1976 and 1978, in rural areas, there has been a sharp decline in all the three fertility indicators i.e. general fertility rate, gross reproduction rate and total fertility rate. About 23 per cent of eligible couples have been effectively protected by contraception. The mid-term evaluation of the sixth plan revealed that 36.7 per cent achievement of planned targets for sterilisation and 30 per cent for intrauterine devices (IUD) resulting in effective couple protection of 25.9 per cent against the target of 36.6 per cent by the end of 1984-85. The performance of different states is not identical. More work remains to be done in states like Uttar Pradesh, Bihar, Rajasthan and Madhya Pradesh.

Ministry of Health and Family Welfare has high targets such as 44 per cent couples to be covered by 1990 and 60 per cent by the year 2000. The average family size of 4.3 (1979) is to be reduced to 2.3 by the year 2000.

Malnutrition

The nutritional status during pregnancy and lactation seriously affects women of rural families. The sex bias in nutrition in favour of the males against females is well established through various nutritional studies (See Table 10).

Table 10
Intake per day of calories (KCal) and iron (mg) in different physiological groups

Physiological groups	Calories (KCal)	RDA KCal	Iron (mg)	RDA (mg/day)
Adult females				
Moderate	1,858	2,000	28.1	32
Pregnant	1,757	2,300	24.3	40
Lactating (moderate)	1,924	2,700	29.6	32
Adult males				
Moderate	2,210	2,400	32.2	24

RDA- Recommended Daily allowance.

Source: Rao B.N. Development of salt fortification programme to prevent iron deficiency in India, National Institute of Nutrition, 1984.

Seeing Table 10, one realises that the intake of calories and iron by Indian women is much below the recommended levels. The result is a combination of overwork and undernutrition. A pregnant woman is unable to gain adequate weight which results not only in low birth weights but also in insufficient fat stores for energy demands of breast feeding. The average birth weight of an Indian infant is estimated as 2.7 kg. Only 5 per cent of Indian infants have a birth weight range of 3.5 to 4 kg.

It is estimated that about 140 million women in India are malnourished. On an average, an Indian woman becomes pregnant eight times, gives birth to 6-7 children of whom 4-6 survive. She spends half of her reproductive life of 30 years in pregnancy and lactation. Her staple diet consists of cereals only. She is deprived of protective foods such as milk, fruits, vegetables and pulses.

Marital Status

A typical factor contributing to high fertility in India is young age at which girls are married. There has been a small rise in age at marriage of girls -- from 13 years in 1901 to 15.3 years in 1951, 17.75 years in 1971 and 18.66 years in 1981. However, the birth rate is high.

It has been also noted that on an average a rural woman gets married two and a half years before her urban counterparts. The Sharda Act prohibiting child marriages (below 15 years) and the amended child Marriage Restraint Act of 1978 (raising age at marriage of girls to 18 years) has not yet been fully implemented especially in rural areas. Many of the rural women are not aware of its existence. Even when they are aware, they prefer to ignore it because of traditions, illiteracy, and fear about safety of unmarried girls.

Marriage is the most valued institution for rural women in India. There is tradition of universal marriage. At age 50, only 5 out of 1000 Indian women remain never married.

It is, however, heartening to see that the percentage of widows decreased from 15.5 per cent in 1961 to 12.5 per cent in 1971. A trend towards a decrease in the incidence of divorce and separation from 1961 to 1971 was also noted.

It has been observed that at All India level, mean age at marriage for female was 18.66 years in the year 1981. Among states, Kerala had the highest average age at marriage of 21.87, while Rajasthan had the lowest mean age at 17.02 years with Bihar closely following with 17.08 and Madhya Pradesh 17.9 and Andhra Pradesh 17.59. Towards the higher side was Punjab with 21.12 years followed by Tamil Nadu with 20.25 years. It is heartening to discover that there is a decrease in the percentage of married females to the total in the 15-19 years age group from 55.41 per cent in 1971 to 43.47 percent in 1981. However, the number of couples per 1000 population on an All India basis remained constant -- 170 in 1971 and 169 in 1981.

The above information about Indian women in general and rural women in particular depicts their profile on some important dimensions. Those interested in studying women can have an idea about the quality of life they are leading in India.

Chapter III

Past Research Work
Radio and Television

In any country, women who are half the population are often half the audience. The success or failure of development plans in education, family planning, community development, health and nutrition depends upon the involvement and participation of women. Since formal education is costly and a long term process, it is essential to harness the mass media for the eradication of illiteracy and speed up spread of basic education among women and girls. The National Council for Women's Education emphasised the potential significance of the mass media to generate public opinion in rural areas in favour of girls' education. The family planning programme is making intensive use of the mass media to inform and create awareness among the people, both men and women. The media have made impact on the audiences which is evident from various studies. However, proper assessment of the present role of media and their impact on rural women would require detailed surveys and studies.

Limited studies have been conducted with media and farm women in India. This has always been a limiting factor in reviewing the relevant literature. Therefore, most of the studies reviewed here have been about the role of various mass media on the behaviour of different sets of individuals particularly in rural set up. The studies reviewed are related to radio, television, newspapers and magazines, films, multimedia combinations, life style and modernization and women's opinions on social issues.(See Chapter IV, V also).

Studies related to Radio

By its reach and impact, the radio provides one of the most powerful media of mass communication. It is specially important as a medium of information and education in a vast country like ours where coverage by the press is not extensive. Available literature having direct or indirect bearing on the present study has been reviewed and presented here.

Profile of Radio Listeners --Age

Sandhu (1970) reported that radio commanded a universal audience in terms of age. Majority of the listeners (55.7%) of the farm radio programmes were, however, within the age group of 31-50 years. Singh (1972) reported that maximum percentage of listeners (56%) were between 32-56 years of age.

Mehta (1972) (LIC 1972) found that age was not significantly related with responses of rural women towards *'gramin mahilaon ka karye-kram'* (a radio broadcast). But, Puri (1972) revealed significant differences with regard to age. She reported younger rural listeners listened to radio more regularly. Bhandari (1972) (LIC 1972) also reported that age influenced the reactions of respondents towards 'religious and folk songs' and 'discussion' programmes of *'Yuv-vani'*. The listeners of older age group preferred religious and folk songs while the younger age group significantly preferred discussions.

In Nepal, Sakya (1973) reported that about eighty two per cent of the listeners of farm broadcasts were adults. Vijayaraghavan (1978) found that members of 'progressive farmer discussion group' were younger (26-29 years) than non-progressive farmer discussion group members (34-37 years) which indicates that younger the listener, better the listening to radio broadcasts. Badrinarayana (1977) reported that the farm-broadcast listening small farmers did not differ in age from non-listeners. A study conducted by Tamil Nadu Agricultural University, Coimbatore (1978) on 'Farm school on the AIR' programme revealed that maximum number of listeners (51 %) were of young age group (less than 30 years). Audience Research Unit of AIR, Delhi (1979) conducted a study on 'Farm school on the AIR' programme and found that over half of the participant farmers were below 30 years. Similarly, Chandrakandan (1981) in Tamil Nadu also reported that most of the farmers registered for 'Farm school on the AIR' programme were young in age. Bhani Ram (1981) found that most of the registered farmers for 'Farm school on the AIR' were young (44%).

Education

As for education of the listeners, Jalihal and Murthy (1974) observed that there was a significant association between formal education of the farmers and their radio listening habits. Dhaliwal and Sohal (1967) found that majority of the radio listeners in Punjab had primary level of education. Sandhu (1970) too observed that radio listener farmers varied among themselves in education and the median education level was primary. Bhandari (1972) (LIC 1972) found that reactions of the rural youth towards 'Yuv-vani' were independent of their literacy level. However, Dhadhal (1973) reported that literate farmers heard the radio rural forum programmes more regularly than the illiterate ones.

Masani (1976) found that it was middle education level farmers who were able to derive more benefits from the radio programmes while the illiterates found the programmes too difficult to follow. Annamalai (1979) reported that education showed a positive and significant correlation with the utilization of farm information sources including radio. Chandrakandan (1981) in Tamil Nadu reported that the farmers registered for 'Farm school on the AIR' programme were highly educated.

Bhani Ram (1981) found educated farmers were able to derive more benefits from the programme 'Farm school on the AIR' probably because they were able to take down notes. A large number of them (55%) were high school pass. Similarly, Talukdar and Pawar (1981) found that education had positive correlation with farmers' perceived utility of farm broadcasts.

Social Participation

Singh and Sandhu (1971) reported that in Punjab, radio listening farmers had high socio-economic status, low social participation, high degree of contacts with extension workers, low contacts with Punjab Agricultural University Scientists and medium participation in extension activities.

Sakya (1973) reported from Nepal that listening behaviour of radio owning farmers was found to be significantly related to closeness with extension workers while that of adult farmers was positively correlated with social participation, closeness with extension workers and participation in extension activities.

According to Soni (1974), only a little less than half of the radio listener farmers included in his study were members of cooperative societies and largest number of the members of 'charcha-mandals' were found to be the members of some formal organisation. Dave (1975) also reported that larger number of the members of farmers' discussion groups were found to be members of some other organisation too. Tamil Nadu Agricultural University (1978) revealed that social participation and contacts with extension agencies were found to influence the listening behaviour of the farmers registered for 'Farm school on the AIR'.

Vijayaraghavan (1978) reported that the progressive farmers' discussion group members were found to differ significantly with respect to social participation indicating thereby that the members of progressive farmer discussion groups had frequent social participation and had higher socio-economic status than the members of non-progressive farmers discussion groups.

Land Holding

Sandhu (1970) reported that in Punjab the audience of farm broadcasts had high socio-economic status and relatively larger farms. Patel (1976) reported from Gujarat that the members of progressive and less progressive farmers discussion groups did not differ significantly with respect to their land holding.

Contrary to this, Vijayraghavan (1978) reported that the progressive farmers discussion group members were having bigger farm size than non-progressive farmers discussion group members wherein progressiveness was the main factor. Tamil Nadu Agricultural University (1978) reported that majority of the 'Farm school on the AIR' had listeners (63%) having small or medium sized farms.

Audience Research Unit (1979) revealed that half of the respondents possessed cultivable land holdings of less than 10 acres. Chandrakandan (1981) found that most of the registered listeners of 'Farm school on the AIR' programme were operators of small and medium farms.

Radio as a Source of Information

Shankariah (1969) found that in a progressive village of Delhi, farmers ranked radio second while in non-progressive villages, farmers ranked it fifth as a source of farm information. Sandhu (1970) also found that radio was ranked third in Punjab as a source of farm information by the farmers. Puri (1972) reported that rural radio programmes were valued because they were a source of agricultural information and people found them helpful in their day-to-day life.

Sakya (1973) in Nepal also showed that radio owning farmers ranked it second while adult farmers ranked it third as a source of information. Ambastha (1974) in his study in Delhi found that radio was at the top as far as its credibility as a source of farm information is concerned. Somasundram (1976) found radio as first in order of preference for communication channels utilised by both adopters and non adopters of farm technology for getting information about all farm practices. Annamalai (1979) also reported that at awareness stage, radio was utilised as a source of farm information for practices like seed treatment and fertilizer application.

So far as farmers discussion groups were concerned, Vijayaraghavan (1978) found that the progressive farmers discussion groups had stronger two-way communication between them and radio, enhancing the process of feedback by sending considerably more number of questions to AIR than non-progressive farmers discussion groups.

Bhani Ram (1981) reported that radio was a source of direct and indirect information as on an average each of the farmer registered for 'Farm school on the AIR' passed on the information to 46 non-listeners. Talukdar and Pawar (1981) reported that majority of the listeners perceived the programmes useful at medium level. Education, cosmopoliteness-localiteness, innovative-proneness, extension activities, annual income had positive correlation with the level of utility of the radio programmes. Mathew (1982), however, reported that effectiveness of the radio for disseminating information and for educating rural women is questionable.

Radio Listening Behaviour

Dhaliwal and Sohal (1967) found that 48.2 per cent farming and 11.5 per cent non-farming families possessed radio-sets. Mehta (1972) (LIC 1972) found that only 64 per cent of the rural women were aware of rural women's programme on radio. Of these, only 39 per cent listened to the programme regularly. All the listeners were satisfied with the afternoon (2.00 p.m.) timings but none with evening timings (6.20 p.m. - 6.40 p.m.), the reason being in the evening the women were busy preparing the evening meals and looking after male members of the family. Duration of programme was acceptable to all. Puri (1972) (LIC 1972) found that only 75 per cent listened to the rural broadcast regularly although all were aware of it. A majority of the listeners (83.3%) were satisfied with the duration of the programme. Bhandari (1972) (LIC 1972) reported that 'Yuv-vani' programme was popular among rural youth as 72 per cent of the respondents considered them 'excellent' and remaining considered them 'good'. Mathew (1982) reported that 20 per cent of the respondents possessed radio-sets although almost all respondents had access to it through neighbours.

Programme Preference

The preference of radio programmes is related to many other variables such as age, sex, occupation etc. Bhandari (1972) (LIC 1972) found that radio is enjoyed more by men-folk. In Delhi villages, Puri (1972) (LIC 1972) found 'Vividh Bharati' (77.5%), Rural broadcast (75%), 'Braj Madhuri' (58.8%) and the 'Hindi News (44.5%) were the most popular radio programmes. So she reported that though the rural listeners were more interested in entertainment programmes, health talks and child-care discussions were also liked by more than 1/6th of the sample. Men showed interest in news and agricultural information. Younger listeners showed more interest in domestic oriented programmes. Women's interests were centred around domestic problems and they looked more for entertainment; whereas men were also interested in information. Farmers liked to listen to agricultural programmes whereas non-agriculturists preferred news broadcasts.

Mehta (1972) (LIC 1972) reported that *'grameen mahilaon ka karya kram'* was enjoyed and appreciated by all groups of women equally. The programmes preferred in rank order were songs, interviews, talks, drama and replies to letters. Sandhu (1970), Singh (1972), Sakya (1973) and Jalihal and Srinivasmurthy (1974) reported that entertainment was preferred by most of the farmers. Masani (1976) reported that radio audience found talks to be one of the least popular programmes.

Mathew (1982) reported that rural mothers' preference towards different radio programmes were film music (70%), 'others'

(15.3%) and 'none in particular' (15.3%). Other programmes of news and farmers' forum were only heard by women who owned radio sets.

Gain and Retention in Knowledge and Attitude Change through Radio.

Radio has been repeatedly indicated in various studies as a medium through which one gains and retains knowledge and changes attitudes. Kishore (1968) found that there was a significant change in knowledge of farmers due to radio broadcasts and also in knowledge retained by them. He also found 'discussion' mode of delivery resulted in more attitude change of the farmers.

Waisanen and Durlak (1968) in Costa Rica also found that changes in knowledge about health and agricultural innovations are more significant for the participants in radio forum group than control group.

Puri (1972) (LIC 1972) found that only 37.5 percent rural people reported gain in knowledge through radio and found the programmes useful in day-to-day life. Mehta (1972) also reported that only 40 per cent of rural women were regular listeners for *'grameen mahilaon ka karya kram'* and also experienced gain in knowledge. Bhandari (1972) (LIC 1972) indicated that audience of 'Yuv-vani' did gain in knowledge through radio programmes. John Knight (1973) reported that maximum gain in knowledge among radio listeners in Tamil Nadu was from interview method.

A study conducted by AIR, Hyderabad (1975) revealed that the broadcast of 'Farm school on the AIR' programme has helped the cultivators in adopting scientific practices of paddy cultivation. Listeners group report of Communication Centre, Tamil Nadu Agricultural University, Coimbatore by Sundarajan *et al* (1978) has indicated that the adodption of technology through radio was 30 to 40 per cent.

So far as adoption of practices by farmers discussion groups and registered farmers of 'Farm school on the AIR' are concerned, Vijayaraghavan (1978) reported that the adoption of farm technology through farmers discussion groups was found to be significantly higher in respect of progressive farmers discussion groups than non-progressive farmers discussion groups. Chandrakandan (1981) also reported a higher adoption of practices through 'Farm school on the AIR' by the registered farmers after listening to the lessons on radio.

Rajamani (1981) studied the impact of farm broadcasts on two organised groups of listeners in Coimbatore district and found that farmers registered for 'Farm school on the AIR' programmes had acquired significantly higher level of knowledge about the improved technology of coconut farming than the knowledge acquired by others. As far as adoption of the technology is concerned, there was no difference between these two groups.

Bhani Ram (1981) found that formal education and social achievement motivation were significantly related with the knowledge of the farm technology through 'Farm schools on the AIR'.

Nigam (1987) (Dte Extn 1987) reviewed findings of the 13 Audience Research studies conducted in different parts of the country. In all, they covered 198 villages and 2500 interviews among rural adult females. Major findings revealed that listenership to radio and the local station was quite high, sometimes bordering cent per cent. Rural women generally tune on to neighbouring stations. They were quite fond of music (film, devotional, light and folk), news, plays, programme for women and children and rural programmes. It was found that around 10 per cent of the rural women were not aware of the programmes meant for them. It was seen that listenership was higher among educated married housewives. In general, the favoured items of women's programmes were family welfare, health and hygiene, mother and child care, social evils, home decor, household hints and cooking. Women respondents were generally satisfied with the frequency, duration and time of broadcast.

STUDIES RELATED TO TELEVISION

The development of television as a medium of communication has immense potential in India. An assessment of the role of television and its impact especially on farm women is, therefore, required before planning future strategies for television communication. A number of studies have been conducted in rural areas to study its effectiveness as a medium of communication.

Socio-personal profile of television viewers --Age

Mishra (1967) found that there was lack of correlation between the age of farmers and their gain in knowledge through television. Thus, television results into gain in knowledge irrespective of age of the farmers. However, the extent of gain varies as Dey (1968) reported that gain in knowledge was more in younger people. Murthy (1969) supported him but, Singh (1969) said that age of farmer had bearing on the selectivity and preference for a communication source. Younger farmers, in general, preferred institutionalized sources and older farmers preferred non-institutionalized sources.

Kaur (1970) reported that highly negative correlation existed between gain in knowledge through television and age of farm women. It was observed that the young farm women gained more knowledge as compared to the middle age and old age women.

Singh (1971), while seeking opinion of farmers on 'Krishi-Darshan' programme found that age had significant influence on the attitude of farmers towards the programmes. Sekhon (1972) found that farmers of all ages showed a strong liking for agricultural telecasts and television programmes as such. Sadamate (1975) reported that the age of the farmers and their viewing behaviour had negative correlation. Jha (1978) found that age of the viewers was positively associated with the duration of television viewing.

Education

Murthy (1969) reported that educated farmers had shown better favourable communication behaviour. Sekhon (1970) studied the effectiveness of television as a medium of communication and revealed that farmers and farm women educated up to matric and above made maximum gain in knowledge. Singh (1970) supported this finding by reporting that education of farmer was related to communication source utilisation.

Kaur (1970) found highly significant positive correlation between formal education of farm women and gain in knowledge through television. Farm women having education upto high school and above gained more knowledge as compared to those with primary school education and the illiterates. Farm women with primary or middle school education also gained more knowledge as compared to illiterates. Sohi (1973) found that educated farm women were more receptive to family planning messages through television.

Sinha (1974), however, reported that formal education was not a limiting factor for television to impart knowledge. Sadamate (1975) found that viewers and non-viewers of 'Krishi-Darshan' differed significantly with regard to their level of education. Ministry of Information and Broadcasting (1981) reported in relation to SITE viewing that illiterates showed more interest than literates in matters such as family planning.

Farm Size, Social Participation and Other Variables

Singh (1969), while summarising researches on agricultural communication, revealed that size of holding had a significant influence on the utilisation of farm information through institutionalised sources. Singh (1970) reported that size of holding and outside contact were the factors associated with the source utilisation. Kaur (1970) found highly significant positive correlation between socio-economic status of the farm women with gain in knowledge through television. Reddy (1971) reported that factors such as farm size and social participation were significantly associated with media exposure and interpersonal communication behaviour. Sekhon (1972), however, found that economic and socio-economic status did not influence television viewing behaviour of farmers. Sohi (1974) found that television messages on family planning were received more favourably by rural women from higher socio-econonic status.

Sadamate (1975) reported that viewers and non-viewers of 'Krishi-Darshan' television programme differed significantly with regard to localite-cosmopoliteness but not with regard to size of holding and social participation. Audience Research Cell (1976) reported that three out of every five television owners also owned a house in Delhi and 35 per cent belonged to joint families. Jha (1978) reported that larger the size of family, higher was the tendency to view TV programmes.

Television Viewing Behaviour

Audience Research Unit of Delhi Doordarshan Kendra (1967) reported that about two thirds of the respondents viewed the programmes daily. It further reported that the viewers consisted of 47 per cent men and 53 per cent women. It further reported in 1973 that 64 per cent of the respondents viewed TV regularly, 18 per cent frequently (4-6 times a week) and 13 per cent occasionally.

In 1976, this Unit reported that on an average 66 per cent or 1.2 lakhs of TV sets used to be tuned daily in Delhi. Average daily viewing was found to be maximum on Sundays (90%) when Hindi feature film was telecast. The viewing was appreciably high on Wednesdays (82%),

Mondays (65%) and Thursdays (64%). On the remaining days it varied from 48 to 58 per cent. On an average TV was viewed for about 44 minutes per day. It was maximum on Sundays (155 minutes) when Hindi feature film was telecast.

In another study conducted by the Audience Research Unit of Madras Doordarshan Kendra (1978), it was reported that 9 out of 10 TV viewers found 6 p.m. time as most preferred for viewing.

Kaur (1970) while studying the farm women televiewers found that television installation plays an important role in television viewing.

Jha (1978) reported that families in Delhi under study kept the TV on for two hours on an average. He also found that possession of TV set for a longer period reduced TV viewing period of the respondents. Wives, in general, started viewing TV programme only after 8 p.m. as they were busy in accomplishing household chores before that.

Programme Preference

Audience Research Unit of Delhi Doordarshan Kendra (1973) reported that 88 per cent of the respondents mentioned 'Chitrahaar' to be their most favourite programme, followed by 'Samachar' and plays. The least preferred programmes were 'Krishi-Darshan', 'Foucs Main', 'Personal-view', 'Topic of discussion' and 'EK drishtikon'.

Again, Audience Research Unit (1976) reported that the feature film in Hindi continued to be the most popular programme of Delhi TV with 87 per cent of the viewers. 'Chitrahaar' came next with 80 per cent viewers. Sports events were fairly popular having 48 per cent viewers.

Similar trends were shown by TV Centre Srinagar (1975) who reported that 48 per cent of the respondents accorded first preference to feature film. Audience Research Unit (Madras) (1978), however, reported that the top preference was given to 'News in English' by more than four-fifth of the respondents. It was followed by play in Tamil and feature film in Tamil.

Pillai and his associates (1974) in their survey on Bombay TV found sports to be more popular among men. Women and children conformed to the universal pattern of being more regular TV viewers.

Sita and Krishnan (1975) reported preference of the programmes of Bombay TV in the following order -- sports, films and film story, science report, Indian and World Affairs, music, drama and children's programmes.

Jha (1978) reported the order of preference for programmes for Delhi TV was -- 'chitrahaar', Hindi feature film, Hindi drama and 'phool khile hein gulshan gulshan'.

Gain in Knowledge, Retention and Influence on Attitude Change

Sinha (1970) found that primary viewers of television gained 50 per cent of knowledge and retained 82 per cent of the knowledge gained. The secondary viewers gained 14 per cent of knowledge and retained 89 per cent of that knowledge after a lapse of 15 days. Kaur (1970) reported that farm women gained 42 to 72 per cent knowledge through television and retained 72 to 83 per cent of the same.

Singh (1971) reported that farmers in television villages were better adopters of package of practices in relation to agriculture. Sekhon (1972) assessed the effectiveness of television on farmers and farm women. She found the gain in knowledge varied from 45 to 62 per cent in the different crops. Sadamate (1975) also supported the view that farmers gained and retained knowledge because of television viewing. Muis (1983) reported that the most striking feature of the impact of television was the substantial increase in knowledge of peasants in Indonesia.

While studying diffusion of information from a farm telecast, Sinha (1973) reported that out of the 42 viewers only 32 could pass on the information to other farmers who did not view television. This supported the theory of two step flow of information. His views were supported by Kamath (1973), Singh (1973).

Starting on Ist August, 1975 and lasting for one year, the Satellite Instructional Television Experiment (SITE) was one of the biggest communication experiments in the world. The ATS-6 beamed instructional programmes in six languages for four hours daily on agriculture, health and primary education to 2,400 villages spread over states of Rajasthan M.P., Orissa, Bihar, A.P. and Karnataka.

The findings indicated the tendency of the viewer to forget caste differences during community viewing. It was also found out that illiterate farmers and village women showed more interest in matters related to family planning etc. Positive role of television was highlighted in educating farmers about agriculture and allied subjects.

Saroj Malik (1987) (Dte. Extn 1987) quoted about an analysis of women programmes originating from Delhi and Jaipur. It showed that maximum telecast time and items were on family planning, health and hygiene, mother and child-care, use of waste materials to prepare decorative items for homes and interviews with known women figures. She also mentioned about problem of erratic and irregular power supply particularly during transmission hours which is often found disturbed and community sets remain idle. According to her conclusion, studies and field observations have shown that majority of rural women do not watch television programmes. The underlying reasons are social inhibition in visiting community sets at odd hours, busy in housework during transmission time and programme not being interesting.

Chapter IV

Studies on Newspapers and Magazines

Due to illiteracy it can be expected that print media have not really reached bulk of farm women. It is, however, encouraging to note that press in India has, by and large, not degraded the status of women. News items highlighting the role of women have been published whenever possible. Some note-worthy achievements by women in different fields appear in daily newspapers and magazines from time to time and some of the major dailies carry a weekly women's section.

Socio-personal Profile of Readers -- Age

Kinyanjui (1972) conducted a survey on the recent development of correspondence education in Kenya. He found that registered learners were in the age group 21-40 years. His findings were supported by Shivpuri (1972) in India who reported that press was the most powerful medium for farm advertisements and was particularly utilised by younger and middle aged farmers.

Khandekar and Mathur (1975) observed that the perceived effectiveness of farm magazines seemed to increase upto the age of 25 years and decline after 45 years.

Education

It is natural that a person who has had more years of formal schooling would have more interest and better understanding of reading material. Khandekar and Mathur (1975) found that education and functional literacy were positively and significantly related to the perceived effectiveness of farm magazine 'Unnat Krishi' published by Directorate of Extension, Govt. of India. Dhillon (1978) also showed education having a significant positive association with comprehension of printed matters in correspondence courses.

Other Variables

Kinyanjui (1972) found that registered trainers of correspondence courses in Kenya owned very few books, did not buy a newspaper regularly but owned a radio set.

Khandekar and Mathur (1975) found that mass media exposure, reading ability, comprehension skill were positively and significantly related to the perceived effectiveness of the farm magazines.

Khajapeer (1978) observed positive and significant relationship between reading comprehension and urban contacts, urban occupational pulls, newspaper reading, radio listening, social aspiration, knowledge and attitude towards adult literacy programme. Fatalism and scientism of the participants were not significantly associated with the level of reading comprehension. Stevens (1980) observed significant relationship between interest and reading comprehension for high ability students. Effect of previous knowledge on the reading comprehension was not observed.

Kaur (1982) found occupation of the family having a bearing on the comprehension of the printed matter. Size of family and its occupation were found to be associated with the extent of use of information by the respondent.

Chatterjee (1973) reported that with 695 daily papers on record, India remained the second largest publisher of daily newspapers in the world. The Union Territory of Delhi has come up remarkably well in the growth of newspapers. The territory put out 29 dailies with a total of 8.17 lakh circulation. There were 21 periodicals in the country with a circulation of above one lakh. The women's magazine 'Femina' was one of them.

A special study of the growth of Hindi and English newspapers in the country during the quinquennium 1964 to 1969 was made by the Registrar of Newspapers. It showed (i) English dailies numbering a little more than one third of the total number of Hindi dailies commanded more than 64 per cent of the total circulation of Hindi dailies (ii) in other periodicals, the difference is less than in the case of dailies, although English newspapers which were 251 fewer than Hindi had

4.27 lakh more circulation, (iii) The percentage increase in number and circulation of dailies of both the languages seem to be even and comparable. In the case of other periodicals, the increase in circulation is proportionately less than the increase in number, while Hindi has shown greater progress. Other periodicals in Hindi show an increase of over 700 in number and in circulation of merely four lakhs.

Indian Institute of Mass Communication conducted National Readership Survey (1978) and reported that 'Sarita', 'Manorama', 'Manohar Kahaniya' and 'Saptahik Hindustan' were among the popular magazines read by Delhi readers.

Schwartz (1981) reported that for most persons newspaper reading is simply another activity, one that if measured in terms of time spent in reading, is a rather minor pastime and did not arise strong reactions.

Printed Matter as Source of Information

Dhillon (1968) reported that farmers of the Punjab state ranked Punjab Agricultural University periodicals at first place as source of information. Schneider (1970) indicated that print media played a positive role in the adoption process particularly at awareness stage. Supe (1971) indicated that written words had positive and significant association with rational behaviour in decision making process of improved practices. It was also observed that written words had positive but non-significant relationship with adoption of improved agricultural practices.

Kinyanjui (1972) reported that correspondence method of teaching can be as effective as conventional face to face teaching. Shivpuri (1972) concluded that press was the most powerful medium for farm advertisements. Newspapers and magazines were ranked at third position in different agro-information media. Manjaiyon (1973) reported that printed media are ranked third in influencing the farmers for adoption of agricultural practices. Khandekar and Mathur (1975) observed that 50 per cent of the readers perceived 'Unnat Krishi' as effective farm magazine.

Sanoria and Singh (1976) found extension publications as one of the most commonly used source of information for all levels of communication. Sohal *et al* (1977) reported that majority of 'Daily-Samachar' readers (60%) found the magazine useful and 47.5 per cent found the information as per their needs.

Kaur (1982) found that majority of women found the lessons useful and liked the content 'Fruits and vegetable preservation' (60.4%), 59 per cent liked 'Food Science' and 56 per cent liked Home Management printed lessons. In a report of the Committee on the Status of Women in India (1974), it was reported about typical youth magazine that they do not provide much food for thought. They imitate their western counterparts and lay emphasis on fashion, music, films and sex. Hassan (1983) reported that for planting of cocoa and coconut 14 and 10 per cent farmers respectively depended on newspapers as source of information.

Chapter V

Studies on Films and Multi Media Approach

Films exert vivid influence on the viewers, specially on social values as well as on mode of behaviour. Films make much wider and deeper impact because of its ready availability and presentation of real life situations.

Socio-Personal Profile of Films Viewers

Indian Institute of Mass Communication (1969) (GOI 1974) reported that youths formed majority of the film goers. It further reported that choice of films varied with their age and sex. Adolescents were the most frequent film goers and out of total sample, 88 per cent viewed films.

Preference Pattern for Films

The same study revealed that categories of films preferred and generally viewed by women were: (i) Films on family life (ii) musicals (iii) devotional films (iv) romantic and (v) tragic films. The least preferred films were on contemporary themes. When asked as to what should be the chief aim and purpose of films produced in India, 42.4 per cent of women emphasized the development of social, cultural and religious values in society.

Chowla (1983) reported that India was the largest producer of cinema films. In 1982, it produced 864 films (now nearly 1000). But the number of regular cinema houses in the entire country is still around

10,000 (now nearing 12,000). There are 229 Field Publicity Units of the Government of India and almost every district in the country has a mobile van with projectors to tour countryside. Most of the feature films produced are on social themes.

Impact of Films on Viewers

A Survey (1957) (GOI, 1974) covering Greater Bombay noticed a marked tendency, particularly among adolescents, to imitate patterns of behaviour shown in the films. Four types of unhealthy influences noticed were in habits of living and spending, manners and mannerisms including fashions in clothes, hair dressing, speech and behaviour towards the opposite sex and immoral and anti-social practices like stealing and prostitution.

National Institute of Audio Visual Education (1961) (GOI, 1974) found that in an experiment, the respondents in the more film group learnt more than non-film group and their retention power was found to be better. Nigam (1964) (GOI, 1974) reported that the amount of factual information learnt was more in the case of pupils learning through film lessons over those learning through class room lessons.

Indian Institute of Mass Communication (1969) (GOI, 1974) reported findings about treatment of sex and love in films. It was seen that 36.4 per cent women regarded it objectionable and 34.5 per cent regarded it unobjectionable.

While reviewing the role and influence of Indian films, Committee on the Status of Women in India (1974) reported that in its content and treatment of women, it lays more emphasis on sex to draw audience. In most social films, woman is invariably assigned a subordinate status in relation to a man and thus continues to perpetuate the traditional notion of a women's inferior status. These films have not attempted to educate women regarding their rights, duties or responsibilities and have ignored reality.

Studies Related to Multi-media Approach

It is rare that a single media is operative in rural situation. Their messages may supplement each other or can be even overlapping. Their total effect is more real and leaves an impact on the life style of farm women. It was, therefore, considered essential to review studies using more than one medium for study for comparative purposes.

Relative Effectiveness of Mass Media as Source of Information

Dhillon (1968) reported that farmers of the Punjab state ranked PAU periodicals at first place followed by extension workers, neighbours and newspapers in terms of their relative use.

Chahil (1972) (LIC, 1972) showed that there was a significant difference in the effectiveness of radio, television and pamphlets in the communication of agricultural information but not for family planning information. For agricultural information, gain in knowledge was highest by television (61.5%), followed by radio (54.4%) and pamphlets (53.8%), Puri (1972) (LIC 1972) found that for effective learning the film plus lecture method was the most effective treatment as compared to lecture only and film only.

Joanne Leslie (1977) reported that a year's programme was conducted by CARE in the Republic of Korea. Nutrition information was broadcast over radio as songs, jingles and brief dramas. At the same time, it supplied calenders, posters, comic books and food charts to radio listeners. The study showed that 53 per cent could state the specific nutrients in each of the food groups.

Another mass media campaign sponsored by CARE, in India by Joanne Leslie (1977) used radio, newspapers, posters, comic books, wall-paintings and other media to carry positive and negative message on the nutritional requirements of pregnant women, on weaning practices and on general nutrition. CARE reported increase of 32 per cent in public awareness with respect to messages related to pregnancy. The percentage of those having knowledge about weaning practices rose from 59 to 93 per cent and awareness of general nutritional matters rose from 72 to 96 per cent.

Joanne Leslie (1977) reported that based on mass media, advertising approach in nutrition education was used in two regions of Equador. The percentage of those who came to value oil had risen from 15 to 74, while those who had understood the nutritional importance of fish and vegetables rose from 0 to 80 per cent. The percentage of mothers actually adopting these supplements increased from 0 to 24 for oil, 17 to 24 for fish and 5 to 15 for vegetables. In another project, he further reported that in two Equadorian Provinces, Imbabma Province and Maubi Province on the coast, radio and posters were used to promote nutrition education over a period of 15 months. Before the broadcasts began, only 30 per cent of those interviewed in Maubi province thought the mother's milk was the best food for infants, while

4.7 per cent believed that fresh cow's milk was the best and 16 per cent thought that powdered milk was most desirable.

Ramson (1977) reported about a project carried out in 500 villages in M.P. in India which employed three types of communication -- mass media, face to face and 9 nutrition rehabilitation programmes. The change in behaviour and attitude of the participants were; (i) increase in awareness of mothers and medical personnel of the need for giving young infants solid foods and (ii) increase of attentiveness among parents of rehabilitated children to the food needs of their young ones.

Joanne Leslie (1977) reported that the housewives association of Trinidad and Tobago (HATT) launched a mass media campaign through radio, newspapers, posters and television. Mothers (85%) could recall advertisement seen on various mass media.

Hassan (1983) reported that 9 per cent farmers in Malaysia depended on radio for information on rubber planting, 2.1 per cent on television and 4.6 per cent on magazines. For information on orchards 9.9 per cent depended on newspapers and 7.3 per cent on radio and television.

Studies Related to Life Style/Modernization

In the preceding studies, it has been repeatedly shown that irrespective of message content, mass media showed an impact in terms of knowledge gained, knowledge retained and attitudes and adoption of new ideas. It is, therefore, natural that combined effect of all these mass media over a period of time will show changes in the life style of people which we may call modernisation. A few studies have shown the relationship of mass media exposure to modernisation and its relationship with other variables.

Lerner (1958) in his classic study of modernisation in the Middle East found that among 300 individuals whom he interviewed in six countries, those who rated high in empathy were also more likely to be literate, urban, mass media users and generally non-tradidtional in their orientation. Lerner's model may be represented as: Urbanization - Literacy - Mass Media Exposure - Income & Voting.

Rogers (1969) used the key concepts of literacy, mass media exposure, cosmopoliteness, fatalism, empathy, innovativeness and achievement motivation in his analysis of the nature of modernisation process. The study was carried out in six Colombian villages. In this study, modernization was viewed as essentially a communication process, modernizing messages must reach the peasant through such communication channels as mass media, change agents or the villagers' trips to cities. These concepts plus literacy (which facilitates media exposure) were considered major antecedent variables in his model of modernization. The main consequent variables were innovativeness, political knowledge and aspirations. Empathy, achievement motivation and fatalism were thought to be intervening variables. The cross-cultural data in his study showed that the best predictors of mass media exposure were functional literacy, formal education and cosmopoliteness. Media exposure was found to be one of the strongest predictors (when included along with nine other variables) of such modernization consequences as innovativeness, achievement motivation and educational aspirations.

Rajamani (1981) reported that progressive and less progressive farmers discussion groups differed significantly with regard to overall modernity. It also found that overall modernity of the members of less progressive farmers discussion group and registered farmers of 'Farm school on the AIR' significantly correlated with the knowledge of technology held by them.

Edeani (1980) reported dominant impact by mass media exposure, interpersonal communication and perceived importance on the process of individual modernity. His findings that the relatively stronger relationship of mass media exposure and interpersonal communication and the far greater impact of interpersonal communication on orientation to change confirmed Roger's contention that the mass media act mainly as sources of factual information, while interpersonal communication is more of a source of direct decision and action.

Schwartz (1981) reported that choice of life style was related to reading interests but 'newspaper' was of little importance to persons in each of the four life style groups identified viz. young optimists, traditional conservatives, progressive conservatives and grim independents. Significant differences between segments appeared on demographic, psycho-graphic and newspaper use variable. However, the most striking was the place of media in the life style of the segments. Chowla (1983) reported that SITE programmes have shown evidence of statistically significant gain in overall modernity of viewers. Muis (1983) indicated that the peasants in post-television era have a higher characteristic of modernisation compared with those in pre-television era.

Studies Related to Women's Opinion on Selected Social Issues

Mass media communication works both ways for reinforcement as well as for change. It is, therefore, expected that women will change and reinforce their views on social issues because of impact of mass media. Some of the studies conducted on women have revealed opinions held by women on important social issues.

Committee on the Status of Women in India (1974) came out with interesting findings. As regards political participation of women, it reported that majority of respondents disagreed that women should vote according to the wishes of the male members and that they should not become members of political party. With regard to women having separate bank accounts, the study revealed that only 25 per cent females had separate bank accounts and 8 per cent had joint accounts mainly with husbands. About widow remarriage, it reported that 62 per cent of females thought it was alright for a widow to remarry. In relation to the issue on childless couples adopting children, 83.2 per cent respondents favoured the idea that widow should be permitted to adopt a child and 51.4 per cent agreed that unmarried woman should be permitted to adopt a child.

The Committee (1974) further reported that 49.5 per cent respondents wanted separate schools for boys and girls at primary and middle level, 60 per cent at high school level and 70 per cent at college level. With regard to children's marriage, the study found that 74 per cent respondents agreed that parents should fix the marriage of their daughters. It also showed that 62 per cent of rural women beyond 60 years of age were widows.

With regard to the issue that men and women should get equal wages, the Committee (1974) reported that 87 per cent of the respondents agreed to it and 58 per cent of respondents agreed with the idea that daughter should have equal share with sons in parental property. Only 13 per cent favoured the idea that men should have freedom to have more than one wife. It also revealed that in 84.1 per cent respondent homes cooking was done entirely by females.

With regard to use of family planning methods, Singh (1972) (LIC 1972) indicated that 70 per cent of rural women were users of birth control measures.

As regards breast feeding, Kaur (1972) revealed that prolonged breast feeding was practised by both urban and rural mothers for 2-3 years. About the use of pure ghee, Aggarwal (1972) reported that buffalo ghee was popular among the higher and middle castes due to prestige values in villages around Delhi. Koshy and Sarin (1970) reported that pure ghee was procured in rural homes through home production because of its high prestige cum health values.

About supplementary feeding Kakar (1972) reported it was given to 56.6 per cent of the urban children before they reached the age of 9 months, whereas in the case of rural children, solid food was given normally after one year. Koshy and Bhagat (1978) also reported that majority of rural women started solids only after the age of 9 months to 1 year.

Khattar (1972) (LIC, 1972) had reported that there was increase in expenditure on social ceremonies because of higher income in farm families because of green revolution. They spent more money on buying of appliances. The greatest increase was in the number of radios, transistors, watches, clocks, fans, electric irons, stoves and heaters.

About the work distribution between mother-in-law and daughter-in-law, Garg (1972) (LIC, 1972) reported that out of 88 household tasks, daughter-in-law was expected to perform 46.9 per cent of the tasks.

Chapter VI

Research Strategy

Impact of various programmes of social change in rural India has been mainly conducted with major emphasis on agriculture and related aspects. However, farm women were rarely made the subject of such studies. Most of the studies on them have been related with their role in agriculture, nutrition, education, family planning etc. There has been a dearth of studies which could focus on the impact of rural development on their life style.

The role of mass media, which is one of the important precursors to modernization, has not been studied enough in the context of farm women. Therefore, this study, which included a new set of variables like life style, mass media exposure and community awareness among farm women could not bank upon any empirical investigations for methodological support. A new set of instruments to measure some of these crucial variables had to be developed to make the research more scientific.

Locale of the Study

The investigation called for an intensive study of the media-use and their impact on farm women which could be advisably undertaken in an area which was comparatively better known to the researchers. The study, therefore, was conducted in three villages of Delhi Territory, namely, Bharthal, Pochanpur and Bamnauli which had been the area of operation of the researchers for nearly a decade. These three villages were also adopted under the Operational Research Project which was being implemented by the Indian Agricultural Research Institute, New Delhi. The investigators had excellent rapport with the village women which was a pre-requisite for the collection of accurate and authentic data. The sample consisted of 336 farm women of which 120 were from Bharthal, 106 from Bamnauli and 100 from Pochanpur. The aim was to have a minimum of 100 farm women from each village to make a fair representation from each village. The details of the selected villages are as follows.

The Villages in the Study

The major crops grown in the area were bajra, wheat, *jowar*, gram and *guar*. Total cropped area in Pochanpur was 235.52 hectare, in Bamnauli 225.50 hectare and in Bharthal 352.12 hectare. Out of this area, 50 to 60 per cent was under high yielding wheat varieties. About 90 per cent of the cultivated land was under assured irrigation and the average size of holding was 1.6 hectare in all the sample villages. The total income from agriculture and non-agricultural sources was Rs. 9028 per family per year. The selected villages were situated 25-30 kms from Delhi. The population of the villages under study ranged from 2000 to 4000. Bharthal was having the largest population with a Senior Secondary School for girls. The other two villages had one middle level school each for girls. All the three villages had a *'panchayat'* and a *'mahila mandal'*, each run by Delhi Administration. Bharthal had a primary health centre. The three villages were within a distance of 3 kms from each other and were well connected by metal roads with frequent (every half an hour) bus service. The average family-size in the three villages was 7.8 with 2.2 males, 2.1 females and 3.5 children.

Female literacy in the three villages is shown in table 11.

Table 11

Female Literacy in the sample villages (%)

Village	Illiterate	Primary	Secondary	College	Total(N)
Bharthal	90.1	6.0	3.6	0.3	648
Bamnauli	89.7	3.8	5.8	0.7	242
Pochanpur	83.9	9.1	7.1	-	298
Total	88.5	6.3	4.9	0.3	1188

Table 11 shows that in the sample villages, 88.5 per cent

women were illiterate, 6.3 per cent were educated upto primary level, 4.9 per cent had passed secondary examination and only 0.3 per cent were college educated. This shows that the level of education in the selected sample villages was low and had more or less identical pattern in all the villages.

Pilot Study

Although data related to family size, education, farming, etc. were easily available, no information on the extent of media-utilization was available. This kind of information was considered crucial as the focus in the study was on media-use and their impact on farm women. A pilot study was, therefore, carried out in the three sample villages to get an overall view of the availability of mass media in the villages. The results of the pilot study were as follows:

Radio was the most common possession among mass media and almost all the families possessed it. Some families even possessed more than one and even upto four radio sets/ transistor sets in their homes. Of course, many of them were out of order. There were about 40 television sets each in the sample villages which were owned by people of all castes and occupations. Commonly subscribed newspapers in the villages were *Navbharat Times* (Hindi), *Hindustan* (Hindi), *The Hindustan Times* (English), *The Times of India* (English), *Pratap* (Urdu), *Milap* (Urdu), *Sandhya Times* (Hindi) and *Evening News* (English). About fifty newspapers were subscribed in each village daily. Magazines subscribed in the villages were '*Sarita*', '*Manohar Kahaniya*', *Soviet Weekly*, '*Grahshobha*', '*Filmi Kaliyan*' and '*Saptahik Hindustan*'. Most of the newspapers and magazines subscribed were in Hindi language. Three cinema halls were commonly visited by villagers which were located 8 to 10 km from the sample villages. A few families in all the three villages also possessed tape recorders and video sets.

Rural women were found to have great interest in entertainment programmes of radio, television and films. Those who could read, took interest in reading magazines and newspapers. Neighbours' television sets were utilized by many women for viewing favourite programmes. Barring a few traditional men, most of them did not object to their wives watching television. Going to films was not very common among women of older age group but younger women did see films off and on. They saw Hindi films only. It was a common practice for newly married couples to go and see films together in nearby cinema halls. Women never went alone to see films, but were mostly accompanied by husbands or some other male members of the family.

The results of the pilot study clearly showed that the village women were definitely exposed to various mass media, although to a varying degree.

Selection of Respondents

A list of farm families in all the three villages was prepared. Women belonging to the age group 15 to 60 years from the farming families were the probable sample respondents for the study. It was decided to take at least 100 respondents randomly from each of the village to give a good representation from each village. A total of 336 women belonging to farming families between the age group of 15-60 years made the final sample for the study.

Variables and Their Measurements

Mass media had been in use in the sample villages for the last several years. The purpose of the study was to have a critical look at the extent of media exposure of farm women and how their life style was being affected.

A careful study of relevant literature indicated that some variables are closely related to mass media exposure and were, therefore, selected for the purpose of the study.

Life Style

This describes the activities, interests and opinions, that, taken together, describe the mode of living of farm women. To study the life style of farm women, a Likert type of scale was developed. The scale constructed was a modified form of the measurement developed by Schwartz (1981) for measuring life style of sub-urban women in the USA. The modifications were mainly done in the indicators/ statements of the scale in order to make them more usable in the Indian context.(See Appendix)

A list of 85 statements depicting life style of farm women covering their home and farm activities, leisure time activities and opinions on common social issues was prepared after reviewing literature, observing farm women and discussing with several women extension experts. This list of statements was mailed to forty experts in the field of agricultural extension, home science extension, adult education and sociology. They were asked to judge: a) whether a particular statement depicted life style, b) if yes, to what degree? i.e., strongly depicted life style, moderately depicted life style, remotely depicted life style or did not depict life style. They were further asked to judge on a 3-point continuum which kind of life style it depicted i.e., whether it depicted i) Modern Life Style or ii) Semi-Modern Life Style or iii) Traditional Life Style. The responses of 25 judges were received.

Forty six statements were finally selected for the purpose of the study. The respondents were finally divided into three categories as per their life style scores i.e., Modern, Semi-Modern and Traditional.

Opinion on Social Issues

Social issues are those which are important from women's point of view, are related to society and which are frequently covered by the media. The issues were mainly socio-economic and related to health and living conditions. Most of them formed an integral part of the set of statements used for studying the life style of farm women. Some other issues which were not covered under the final life style items, were asked in the form of open ended questions in the interview schedule especially developed for the study.

Innovativeness

Innovativeness is the degree to which an individual is relatively earlier to adopt new ideas or innovations. Rogers (1969) predicted that innovativeness and mass media exposure influenced each other. It was, therefore, thought important to take it as a variable for the present study. Innovativeness scale developed by Feaster (1968) was used to measure this variable. The scale consisted of 12 items out of which 8 were positive and 4 were negative. Scoring for different items was as follows:

Responses	Strongly agree	Agree	Undecided	Disagree	Strongly disagree
Positive Item Scores	5	4	3	2	1
Negative Item Scores	1	2	3	4	5

Following the above scoring procedure, responses for each item were scored separately. Scores for all the 12 items were, then, summed up to give the total innovativeness score for an individual.

The respondents were divided into three categories i.e. Low Innovative group, Medium Innovative group and High Innovative group.

Community Awareness

This is the extent of knowledge farm women possess about on-going programmes of community interest in the village.

Interview schedule was developed to study community awareness of programmes going on in the community such as that on education, health, factories, banking facilities etc. Farm women were asked to find out awareness about community programmes running in their village on 12 items. One score each was given for awareness about one programme. Maximum obtainable score was 12. Farm women were categorised in three categories according to their scores: Low community awareness, Medium community awareness and High community awareness.

Mass Media Exposure

Mass media exposure is the extent to which a farm woman is exposed to different mass media of communication such as radio, television, newspapers, films etc.

To develop the mass media exposure scale, the help of expert judges (same as those used for selection of items for studying life style) was sought. Each judge was asked to divide a total score of 100 to the four mass media items namely, newspapers and magazines, television, radio and films. The total scores for newspapers and magazines were further to be subdivided between newspapers alone and magazines. The scores for newspaper were to be further split between daily, weekly and fortnightly newspapers. The scores for magazines were to be split for weekly, fortnightly and monthly magazines.

The judges were also asked to give relative weightage of their scores given for television, radio and films for daily, weekly, monthly and "rare" exposures. The responses of judges were tabulated and an average mean for the responses was found. The scale so developed was tested for its reliability and validity. The respondents were divided into three categories according to mass media exposure i.e., High mass media exposure group, Medium mass media exposure group and Low mass media exposure group.

The other variables included in the study were, age, education, socio-economic status, extension contact, localite-cosmopolite, scientism-fatalism, type and size of family.

Collection of Data and Analysis

Looking into the objectives, a list of information to be collected was prepared. Then, an interview schedule was developed for collecting data from farm women. The schedule mainly consisted of the following parts.

1. Background information.

2. Instruments developed or taken from other research studies for collecting information on independent and dependent variables.

3. Information on mass media utilisation.

4. Information on opinions of farm women about the impact of selected media on their life style.

5. Information on opinions of farm women on selected social issues.

The schedule was pretested on 30 farm women in a nearby village *Bijwasan*. The interview schedule was suitably modified as per pre-testing experience. Chi-square, percentage, correlation, multiple correlation and regression analyses were used to analyse the data.

Micro Level Profile of Farm Women

The present study was conducted in Delhi villages. Out of the total 336 respondents, majority (132) belonged to young age group (15-24 years) while 114 belonged to middle age group (25-45 years) and 90 to old age group (46-60 years).

Table 12

Distribution of farm women according to their age

Age	Modern N (%)	Semi Modern N (%)	Traditional N (%)	Total N
Young	38 (28.8)	94 (71.2)	-	132
Middle	36 (31.5)	72 (63.2)	6 (5.3)	114
Old	-	26 (28.9)	64 (71.1)	90
Total	74 (22.1)	192 (57.1)	70 (20.8)	336

It can be seen that those with modern life style belonged to young and middle age groups only. This indicated that majority of the old age group had developed a set pattern of traditional life style. Similar findings about relationship between age and modernity have been earlier established by Lerner (1963), Rogers (1969), Mcnelly (1970) and Edeani (1980).

The farm women were divided into three categories viz. low, medium and high, according to the extent of exposure to mass media. Table 13 shows the distribution of farm women according to mass media exposure and nature of life style.

Table 13

Distribution of farm women according to their mass media exposure

Mass media Exposure	Modern N (%)	Semi-modern N (%)	Traditional N (%)	Total N
Low	-	38 (45.2)	46 (54.8)	84
Medium	28 (17.3)	112 (69.1)	22 (13.6)	162
High	46 (51.1)	42 (46.7)	2 (2.2)	90
Total	74 (22.1)	192 (57.1)	70 (20.8)	336

About 25 per cent of women had low mass media exposure whereas 26 per cent had high mass media exposure and the rest were categorised as having medium mass media exposure. Similarly, about 22 per cent farm women had modern lifestyle while 57 per cent had semi-modern and 21 per cent had traditional life style. It was revealed that more than 62 per cent of the modern women had high mass media exposure while none of them had it low. In the semi-modern category about 58 per cent farm women had medium mass media exposure with about 20 per cent having high exposure. However, among the traditional life style category, 65 per cent farm women had low mass media exposure and over 30 per cent had medium mass media exposure. Life style had very high positive correlation with mass media exposure ($r=0.681$).

Innovativeness is another important variable which is closely linked with life style. Table 14 shows the relationship between innovativeness and life style which was discovered as highly significant.

Table 14

Distribution of farm women according to their innovativeness

Innovativeness	Modern N (%)	Semi-modern N (%)	Traditional N (%)	Total N
Low	-	4 (6.9)	54 (9.7)	58
Medium	22 (15.3)	108 (75.0)	14 (9.7)	144
High	52 (38.8)	80 (59.7)	2 (1.5)	134
Total	74 (22.1)	192 (57.1)	70 (20.8)	336

Most of the traditional women (93.1%) were found to be less innovative, majority of semi-modern farm women (75%) had medium level of innovativeness. Within the high innovative group, semi-traditional were largest in number followed by modern life style group. Traditional farm women (1.5%) having high innovativeness were negligible. Rogers (1969), Mcnelly (1970) and Edeani (1980) also reported similar findings.

Community awareness is an important factor related to life style. The farm women were asked about the on-going community programmes in their village. They were divided into three categories based on their awareness as shown in Table 15.

Farm women were generally up to date about the on-going community programmes in their village. Majority (218) out of 336 had high community awareness. The farm women came to know about these programmes mainly through interpersonal sources of communication. Mass media did not give area specific coverage on such programmes.

Table 15

Distribution of farm women according to their community awareness

Community awareness	Modern N (%)	semi-modern N (%)	Traditional N (%)	Total N
High	39 (17.8)	121 (55.5)	58 (26.6)	218
Medium	11 (18.3)	39 (6.5)	10 (16.7)	60
Low	24 (41.4)	32 (55.2)	2 (3.4)	58
Total	74 (22.1)	192 (57.1)	70 (20.8)	336

Education was found to be closely associated with life style (see table 16).

Table 16

Distribution of farm women according to their education level

Level of Education	Modern N (%)	Semi-modern N (%)	Traditional N (%)	Total N
Very low	-	46 (43.4)	60 (56.6)	106
Low	2 (8.3)	20 (83.4)	2 (8.3)	34
Medium	72 (34.9)	132 (64.2)	2 (0.9)	206
Total	74 (22.1)	192 (57.1)	70 (20.8)	336

It was discovered that none of the modern farm women had very low education. Most of the traditional farm women (62.3%) were in very low education category followed by semi-traditional farm women (43.4%) Majority of the farm women (83.4 %) in the category of low education were semi- modern. Farm women having medium level of education belonged either to semi-modern category (64.2%) or modern category (34.9%). An overall picture showed that farm women generally had medium level of education. It was seen that women with higher education were more modern. Similar findings have been reported by Lerner (1963), Rogers (1969) Mcnelly (1970), Committee on the status of women in India (1974), Edeani (1981) and Schwartz (1981). It was described as one of the most important factors related to modernisation of an individual.

Chapter VII

Radio
The Most Powerful Medium

The potential of radio in promoting development in Indian rural setting is far greater than any other tool of mass media. By its reach and impact, the radio provides one of the most powerful media of communication. It is especially important as medium of information and education in a vast country like ours where coverage by the press is either not extensive or is limited to certain sections of the population. Reaching people through 98 radio stations, 182 transmitters and four FM transmitters, it covers 95 per cent of the Indian population and 86 per cent of the area of the country(Manorama Year Book, 1988). A large expansion has been planned for the Seventh Plan period when the country will have 204 transmitters and 104 FM transmitters. By the year 1989, radio signals are expected to cover 100 per cent of Indian territory.

There are, however, only 4.4 radio sets per 100 persons in India(Chowla, 1983) as against UNESCO recommendation of minimum five radio sets per 100 persons.

Radio - An Entertainer

One of the manifest functions performed by the radio is that of an entertainer. It brackets the day of the individuals from morning till night and fills up the leisure moments with various types of entertainments including music, plays etc. However, the latent functions performed by the radio are even more important than manifest functions. These include imparting knowledge, helping in framing opinions -- socio-economic, political etc. Besides a family entertainer, it has become an important source of information in relation to culture and values. Amongst the mass media, radio has become a household medium and has entered many of the households as part of daily routine. Peripheral listening of radio while performing other activities is a common feature in Indian homes. The development content of radio is an important source of awareness about citizenship, health and hygiene, education, culture and many other aspects of life.

Radio and Farm women

Radio serves as an important link between important 'happenings' and the millions of women in India, majority of whom live in villages. Radio in some cases is their only link with the outside world. Realising the urgency and importance of reaching farm women, fairly good coverage to programmes of their liking is being given. Their literacy percentage being only about 13, they get informal education on diverse subjects like child rearing, home management, domestic hygiene, home economy etc. The rural broadcasts carry practical information on rural life to increase their knowledge and understanding of national and international events. Some women look forward to solutions suggested on radio for solving personal and family problems. Fortunately, radio has potential to reach even an average farm woman since there is no stigma attached to radio listening even in rural cultural set-up. By and large, the programmes for rural women are put out one or two times in a week. Majority of the stations put these programmes during noon between 12.30 and 2.00 p.m. Generally the duration of the programme is 30 minutes.

Radio Owning by Farm Women

Owning radio in rural area is relatively a recent phenomenon. In villages under study in Delhi, it was found that 86 per cent of the radio owning families purchased them in the last 10 years(About 42 per cent in the last 5 years). However, about 14 per cent farm families did not own a radio set. Interestingly, all the farm women with modern life style had a radio.

The farm women have expressed several reasons for owning a radio set in the family. The reasons were closely related to the functions performed by radio in rural set up. The following Table 17 shows the reasons given by farm women for owning a radio set.

Table 17

Reasons given by farm women of different life styles for owning radio sets.

Reasons	Modern (N=74)	Semi-modern (N=192)	Traditional (N=70)	Total (N=336)
Entertainment	54	78	86	74
News	81	89	14	71
Subject matter knowledge/education	74	78	7	63
Pressure from family members	20	26	86	34
Gift in dowry	7	16	79	27
Any other	-	4	3	3

Multiple Responses.

A majority of the farm women(74 %) expressed they owned a radio set because it 'entertained' them. This was closely followed by other reasons like 'news'(71 %) 'subject matter knowledge/education' (63 %), 'pressure from family members'(34 %) and 'gift in dowry'(27 %). It can, therefore, be safely said that radio was identified by farm women as an entertainer-cum-informer.

It was found that 86 per cent traditional women reported radio as an 'entertainer' as compared to 78 per cent semi-modern and 54 per cent modern farm women. However, the 'news' reason for radio ownership was mentioned more by semi-modern (89 %) and modern (81 %) farm women. Similar trend was seen for the reason 'subject matter knowledge/education'. This trend is quite understandable. Radio is now the medium of information and education for semi-modern women whereas for traditional women it is still more of an entertainer. It is likely that modern farm women have switched over to other media like television and printed materials as sources of information and news. The reasons such as 'pressure from family members' and 'gift in dowry' were mentioned mostly by traditional farm women. This showed that traditional women did not always have radio at home because of their own choice but got it either because of 'pressure from family members' or received it as 'gift in dowry'. It seems radio has become a common dowry item in farm homes.

Programme Preferences

Programme preference is a changing phenomenon. However, it is a reflection of farm women's reasons for owning a radio set and also reflects their life style. Since entertainment is the dominant reason for owning a radio, the listeners have preference for programmes which provide entertainment. Farm women with modern life style have different patterns of programme preferences. This does not mean that they do not prefer entertainment on radio but their inclination is also to listen to serious programmes like news, programmes on health, political lectures, group discussion etc. A typical programme preference profile of farm women in Delhi villages is presented in the following Table 18.

Table 18

Farm women's favourite radio programmes (%)

Favourite radio programmes	Modern (N=74)	Semi-modern (N=192)	Traditional (N=70)	Total (N=336)
Women's programmes	66	67	66	66
News	79	70	40	66
Drama	58	74	49	65
Film songs	66	71	37	63
Rural programmes	55	63	66	62
Religious songs (Bhajans)	32	47	43	43
Health	42	37	31	37
Sports	24	39	14	30
Political lectures	32	37	11	30
Advertisement/sponsored programmes	26	19	9	18
Group discussion	16	15	17	15
Any other	16	7	-	8

Multiple Responses.

The women's programmes which are generally educational in nature were most popular (66 %) followed by news (66 %), drama (65%), film songs (63%) and rural broadcasts (62%). It was a positive trend since except for film songs, rest of the programmes are educational in nature.

Women's programmes were equally popular amongst the farm women of all the life style categories. News was most popular among modern (79%) farm women followed by semi-modern (70%) and traditional farm women (40%). Film songs happened to be most popular among semi-modern (71%) rather than modern (60%) and traditional farm women (37%). It was attributed to the fact that modern women sought more information from radio than entertainment. Rural programmes were most popular among traditional women (66%) followed by semi-modern (63%) and modern farm women (55%). This means that rural programmes were effective because of their correct cultural tilt. 'Bhajans' (religious songs) were another popular programme among semi-modern (47%) and traditional farm women (43%). It shows that limited number of modern farm women (32%) took keen interest in religious songs broadcast by the All India Radio. It was noted that a large number of modern farm women (42%) took interest in health related programmes. This shows as a farm woman becomes modern, she takes more interest in betterment of family health.

Interestingly, the educational programmes fortified with entertainment have become equally popular among farm women.

A large majority of farm women declared that women's programmes on radio were their favourite radio programmes. Puri (1972) (LIC, 1972) also reported that large number of rural women in her sample showed interest in programmes centred around domestic problems. Mehta (1972) (LIC, 1972) reported that *'grameen mahilaon ka karyakram'* was enjoyed and appreciated by all groups of women equally.

The other programmes preferred by farm women were news, drama, film songs, rural programmes, religious songs and health programmes. Puri(1972) (LIC,1972) also reported similar findings i.e. 'Hindi news', 'Vividh Bharati'(film songs), radio rural broadcasts, 'Braj Madhuri' were some of the favourite programmes of rural people. She also reported that though the rural listeners' interests were more toward entertainment programmes, talks related to health and child-care discussions were also liked by 1/6 of the sample. Mehta(1972) (LIC,1972) also found that programmes preferred by rural women in order were songs, interviews, talks, drama and replies to letters. Bhandari(1972) (LIC, 1972) reported that among 'Yuv-vani' rural audience the most popular programmes were the news, Indian music and replies to letters. The older age group preferred religious and folk songs and younger discussion programmes. Mathew(1982) reported that rural mothers' preference for radio programmes was film music. She, however, reported that none of the rural mothers showed interest in educational programmes. This is contrary to the findings of the present study, i.e. women's programmes and rural programmes which are educational in nature were preferred by farm women. This could be because of longer contact of farm women with radio broadcasts in selected villages. It can, therefore, be concluded that the farm women generally liked radio programmes in this order of preference: women's programmes, news, drama, film songs, and rural programmes.

Impact of Radio

The impact of mass media on people is a subjective concept. They have visible influences on their life which are difficult to analyse. One of the major impacts of mass media is the change they introduce in the life style of the users which is the main assumption of several studies in the area of mass communication. The audience can, however, express their opinions on the possible impact of media on their life style. The following Table 19 lists the opinions expressed by farm women in Delhi villages in order of importance:

The farm women opined that radio provided education to them for improving their living, increasing their knowledge and providing information on home improvement. This clearly indicated that radio was an important source of education and a part of their life.

It added to their knowledge and also 'entertained and gave happiness' as mentioned by 59 per cent respondents. Another important impact of radio was that it 'gave knowledge on social issues' (56%), provided 'education for health' (54%) and 'knowledge on agriculture' (54%). It was observed that most of the modern women (87%) felt that radio provided 'education for better living' followed by Semi-modern (63%) and traditional farm women (63%). The same trend existed for 'increase in knowledge on social issues'. Traditional group (66%) considered radio more of an entertainer followed by modern (63%) and Semi-modern farm women (55%). The largest number of modern farm women (68%) found radio listening having impact on health followed by Semi-modern (54%) and traditional farm women (37%).

Many studies have reported similar findings about the impact of radio as a medium through which one gains knowledge. Kishore(1968), Kivlin et al(1968), Waismen and Durlak(1968) all reported that farmers gained knowledge through radio on agriculture and health related subjects. Puri(1972) (LIC, 1972) found that only 37 per cent rural people reported gain in knowledge through radio and found the programmes useful in their day-to-day life. Mehta(1972) (LIC, 1972) also reported similar finding that only 40 per cent of rural women were regular listeners of *'grameen mehilaon ka karayakram'* and gained knowledge through it. The reasons for higher percentage (68%) of farm women reporting gain in knowledge through radio in present study as against studies by Puri(1972) (LIC, 1972) could be that in last ten years radio has spread more rapidly in Delhi villages. It could also be that because of higher number of educated women in the sample of the present study, higher gain in knowledge was reported. It can be said that impact of radio on the life style of rural women was dominantly educational particularly for modern farm women.

Table 19

Farm women's opinion on impact of radio listening on their life style (%)

Opinion	Modern (N=74)	Semi-Modern (N=192)	Traditional (N=70)	Total (N=336)
Education for better living	87	63	63	68
Increases knowledge	84	66	54	68
Information on home improvements	76	60	57	63
Entertains and gives happiness	63	55	66	59
Knowledge on social issues	79	53	40	56
Education on health	52	54	37	54
Knowledge about agriculture	66	51	49	54
News and information	58	55	34	51
News from foreign countries	50	44	31	43
Bad influence of film songs	5	4	6	5
Any other	5	2	3	3

Multiple Responses

Chapter VIII

Television - Medium of the Future

The recent growth of television in India and its increased use in entertainment and education has been phenomenal and unprecedented in the history of mass media. In India, after making a rudimentary beginning in 1959, the regular telecast began on the Independence Day in 1965. For the next seven years, there remained only one T.V. centre i.e. in Delhi. In fact, the expansion had been more rapid in the late seventees and the eightees.

There are18 TV centres with production and relay facilities and in all 274 transmission centres in the country(December,1988). By the end of Seventh Five Year Plan(1990), it has been proposed to have more studios with programme production facilities and more than 400 T.V. transmission centres. At that time, T.V. will cover 90 per cent of the population of the country. Though the TV signals shall reach to such a vast population, only 10 per cent shall be able to see them because of less availability of TV sets.

The TV Dilemma

Women and children form a major chunk of TV viewers (75%) in our country. They are directly affected by the content of the TV programmes, perhaps because of their longer exposure to TV. However, the managers face the dilemma of choosing appropriate programmes to make a positive impact on them. The main dilemma is whether to use TV for information and education or entertainment or all rolled up into one. Such an issue would attract a lot of academic discussion even with political overtones. This would also be a difficult exercise for managers of the software. TV being a state medium has certain functions to perform which fit into the overall media policy of the state. However, irrespective of the functions performed, TV influences lifestyle of the viewers.

Realizing that woman is a separate social entity, having her own information needs, Doordarshan has alloted a separate portion for her in its transmission schedule. e.g. Delhi Doordarshan has three programmes for women in its weekly menu - 'Ghar Bahar', 'Mahilaon Ke Liye' and 'Grameen Mahilaon Ke Liye', each for a duration of about 30 minutes. The last one is for rural women.

TV Ownership

TV set is no more merely a status symbol in Indian society. In urban middle class families, it has become an essential household item. However, in rural society, some prestige is still attached to TV ownership.

In spite of the recent growth of TV industry, the TV ownership is still less than the desirable level.

Possessing a television set is a pre-requisite for regular viewing. In the study conducted in Delhi villages, it was observed that television was not commonly owned by farm families. Out of 336 farm families studied, about 30 per cent owned it. Of the remaining farm families, about 28 per cent viewed it at neighbour's place. Rest of the non-owners did not like to view TV even at neighbours. Table 20 gives the details. It clearly shows that majority of the sets were installed during the period 1980-83.

Table 20

Television ownership in Delhi villages according to life style categories(%)

	Modern (N=74)	Semi-Modern (N=192)	Traditional (N=70)	Total (N=336)
Do not own T.V.	53	72	80	70
Owned for less than a year	3	9	11	8
Owned for 1-3 Yrs.	22	10	6	12
Owned for more than 3 yrs.	22	9	3	10
	100	100	100	100

Owning a television set or viewing it at neighbours was more common among modern farm women sharing their keenness for TV programmes. Therefore, the TV ownership in Delhi villages was a recent phenomenon and was gradually picking up. This is so in spite of TV making appearance as back as 1965.

Entertainment (84%) seems to be the dominant reason for owning a television. This reason was closely followed by 'to hear news' (79%) and to 'gain knowledge' (76%). Some families got it as a 'gift in dowry' (41%) or purchased it because of 'pressure from the family members' (33%). A few families (20%) purchased it because it was a status symbol.

Non-owners, however, could not afford to buy a television set due to its 'prohibitive price' (74%). Many families rationalised not owning a set by saying 'they do not have time to watch television' (60%) or that it was 'harmful to the eyes' (55%). Many elders considered television a 'source of negative influence on the younger generation' (43%). Some (12%) enterprising farm families were waiting to receive a set in dowry. Why to spend money on a television set when neighbours allow them to view it without any problem (3%).

Most of the non-owners who were non-viewers also, did have TV sets at their neighbours. However, about 68 per cent of the non-owners did not view TV even at neighbours because the owners did not encourage this practice. Some of the non-owning non-viewers themselves did not like to view TV at others' place (53%). Other reasons for their non-viewing were as follows:

(i) Programmes were not useful (53%);
(ii) No time to view TV (43%);
(iii) Below dignity to view TV at neighbours (40%);
(iv) TV located away from home (23%); and
(v) Programme timings not suitable (17%).

Obviously, the tradition of free intermixing with neighbours seems to be gradually disappearing and people do value their privacy for matters such as viewing television. After all, it has become a personal medium and has already moved from drawing room to bedroom.

Programme Preference

Since entertainment and recreation were the major motives for owning a television set, the programme preferences centred round these motives. Shahi(1977) reported that recreation was the prime reason for television viewing. Audience Research Unit(1973, 1976), T.V. Centre, Srinagar (1977) and Jha(1978) have all reported that entertainment was the prime objective of television viewers. Other important reason was to get news and information. Audience Research Unit(1973) reported that news was the second most popular television programme. Audience Research Unit, Madras(1976) reported that 'News in English' was given top preference by four-fifth of the viewer respondents. The third important reason for getting television installed at home was to get subject matter knowledge or education. Many research studies have indicated that television was considered the main source of information. Shahi(1977) who studied programme preferences in SITE, reported education as one of the important reasons for viewing television. Prior to these studies, Sinha(1970), Kaur (1970), Singh(1971), Sekhon (1972) and Sinha(1973) had all reported that television was a source of information through which rural people gained knowledge.

The suitability of television programmes for farm women is a question which needs to be researched. Whether farm women get enough time to view programmes of their choice or otherwise also calls for further probing even among farm women. This has specially happened in the last two to three years. Obviously, the viewers among farm-women have developed certain preferences for different programmes. Table 21 gives the details:

Feature films were preferred the most followed by film songs and drama. Other popular programmes were folk songs, news and Krishi Darshan. Feature films were liked more by modern farm women and the least by traditional farm women. Similar trends were seen for film songs and drama. The modern farm women had more interest in agricultural programmes than others. Surprisingly, women's programmes on television were much less popular amongst women. This indicated that either there was a dearth of women-oriented programmes or the existing programmes did not cater to farm women's information needs.

Since programme preference is a dynamic concept, there is a need for making regular assessment and modify programmes accordingly.

Impact of T.V.

Is there any impact of television on farm women? There is no

Table 21

T.V. programme preferences expressed by farm women of different life styles (%)

Preference	Modern (N=74)	Semi-modern (N=192)	Traditional (N=70)	Total (N=336)
Feature film	73	47	34	50
Songs of feature films	67	42	20	43
Drama	49	31	17	32
Folk songs	24	12	20	17
Sunday entertainment	13	5	3	6
Interview of film stars	5	7	3	6
Agriculture prog.	21	14	14	15
News	8	19	11	15
Women's prog.	1	5	4	4
Children's prog.	16	9	6	10
Sports	5	5	-	4
Ads.	3	3	-	3
Any other	11	1	-	3

sure method to know this subjective aspect of change in knowledge, behaviour, attitude and opinions of farm women. It is only possible to know what they express verbally. However, television exerts powerful influence on whatever viewers do and think. This is why the ownership of this medium is frequently debated. Television through which one gained and retained knowledge was reported by Sinha(1970), Kaur(1970), Singh(1973), Sadamate(1975) and Shahi(1977).

Data collected in Delhi villages brought out an interesting observation that while buying television, entertainment was the main consideration. Soon after, the farm women discovered its educational value as well. The detailed observations have been given in Table 22.

Table 22

Farm women's opinion about the impact of television on their life style(%)

Opinion	Modern (N=74)	Semi-Modern (N=192)	Traditional (N=70)	Total (N=336)
Knowledge about:				
New developments	89	68	40	67
Women related matters	67	55	14	49
Home crafts	64	49	14	45
Agriculture	56	37	11	36
Home practices	54	35	14	35
Sports	51	24	8	26
Children's problems	5	7	2	6
New products	8	2	1	6
Social issues	5	7	-	5
Improve mental level	89	70	37	67
Entertainment				
Entertains and gives happiness	67	49	14	45
Story-teller & poem reciter	51	21	8	25
Leads to better human relations	8	8	5	7
Bad influence on women	8	15	65	24
Information				
In touch with urban life.	56	37	8	35
Awareness about current prices.	51	21	8	29
Others	5	2	2	3

Multiple Responses.

Majority of the farm women realised that TV improved mental level, provided knowledge on new developments, especially about women related matters. Therefore, impact of television as a medium of information and entertainment was realised more by modern rather than traditional farm women. One aspect which needs attention is the importance attached by modern farm women to depend on television for keeping in touch with urban life. This is a testimony of two facts: (i) Urban culture attracts farm women, and (ii) programme content on television having urban bias attracts the attention of farm women.

T.V., although, was a source of information and entertainment for farm women, a large number of them could not own it because of financial limitations.

The idea of group viewing was not acceptable to most of the farm women since they were even hesistant to view it at neighbours. Therefore, community viewing of television by farm women need not be encouraged. Ownership of private television sets only need to be encouraged.

The initial ownership of television was mainly because of entertainment motive. Farm women would welcome more of interesting and entertaining programmes. Therefore, programmes meant for farm women should be entertaining with subtle developmental messages. It is only then, that programmes like Krishi Darshan would pick up with farm women.

The study indicated that agricultural programmes and women's programmes were very low in preference pattern of farm women on choice of television programmes. A lot needs to be done regarding this. As far as agricultural programmes were concerned, their coverage was found adequate but their content and treatment were not up to the mark. These programmes need to be presented in interesting and simple manner, so that farm women find them beneficial and get motivated to view them regularly. Farm women are equal partner in performance of farm operations and their exposure to modern agricultural technology is as essential as for farmers. Coverage of farm women's programmes on television was found inadequate. Most of the women's programmes on television are meant for urban women and that is why farm women found them irrelevant. Existing women's programmes need to be presented keeping in mind needs, interests and background of farm women. These programmes should help them develop positive views on important social issues like dowry system, family planning, women's education etc.

Impact of television on farm women's lifestyle was generally considered positive. It means television was well received by farm women, and it can really be used as an educational tool to enlighten the minds of illiterate farm women. This is a happy beginning.

Chapter IX

Print Media - a Limited Reach

Despite enormous expansion of primary and adult education in the last forty years, the literacy rate in India is only about 36 per cent (Census, 1981). The capacity to read newspapers and other written materials with ease in any language would not be more than 23 per cent (Sarkar, 1982). Literacy among females has been even much less than males. About 47 per cent males were literate in 1981, as against only 23 per cent females. Literacy rates for males and females in rural areas were 47 per cent and 18 per cent respectively. About 38 per cent of rural families of India which reside in Bihar, Madhya Pradesh and Uttar Pradesh had rural female literacy of below 10 per cent. Low literacy and poor purchasing power are partly responsible for the limited reach of newspapers and magazines in rural areas. Though circulation of newspapers in India is as low as 2 copies per one hundred persons, readership per copy is much higher than the developed countries. However, in spite of illiteracy, Press has become a major institution of change in India.

Exposure to print media has been playing a crucial role in information. About 19,937 newspapers and other periodicals hit the news stands of which 1334 are dailies, 5701 weeklies and the rest are other periodicals. In the last two decades from 1963, the number of dailies increased by nearly 183 per cent.

Language-wise distribution shows that 5,655 and 3,689 newspapers were published in Hindi and English respectively. These were followed by 1,463 in Bengali, 1,293 in Urdu, 1,234 in Telugu and 1,048 in other languages. Hindi newspapers had the highest circulation of 13,984 thousand followed by English at 11,039 thousand. Combined circulation of Indian Press has crossed 52 million. In 1983 there were

192 big, 427 medium and 7356 small newspapers in the country. The Press in Delhi maintained top position in circulation with 81.76 lakh copies in 1983. Delhi also has the distinction of publishing 15 out of 16 newspapers in principal languages (Press in India, 1984).

Though the circulation figures seem impressive, the fact remains that nearly 93 per cent of the total sale of daily newspapers was confined to the large urban areas where 10 per cent of the people live (Yadav 1983). This means newspapers have yet to penetrate the rural areas. In all development programmes, the supportive role of press can hardly be denied. How far this role is being performed is of interest to all concerned with rural development. Since women are equal partners in development, influence of mass media on them has acquired added interest. Gradual popularity of print-media among farm women makes one wonder how they feel about its effect on them.

The situation in villages under study presents a somewhat different picture. Several daily newspapers and magazines found their way amongst the farmers and farm women of selected three villages. However, women mostly read Hindi newspapers and magazines. Some of the commonly subscribed newspapers in the villages were *Navbharat Times* (Hindi), *Hindustan* (Hindi), *The Hindustan Times* (English), *Employment News* (English), The *Milap* (Urdu), *Sandhya Times* (Hindi) and *Evening News* (English). About 50 copies of newspapers daily come to each of the villages.

Magazines subscribed were '*Sarita*', '*Manohar Kahaniya*', '*Soviet Weekly*', '*Grah-Shobha*', '*Filmi Kaliyan*' and '*Sapthaik Hindustan*' (all in Hindi). Several other magazines were read occasionally.

On the whole, up to 40 per cent of the farm women were reading or hearing news from newspapers. Obviously, the main reason is that the newspapers/magazines are readily available in the villages around Delhi. Moreover, the farm-women get opportunity to read or hear newspapers etc. which are generally subscribed by the male members of the family some of whom have high urban contacts.

Exposure to mass media is a well recognised precursor to modernization. That newspaper readership influences lifestyle of individuals has been established by Schwartz (1981). In Delhi villages, a large number of rural women were found to read or hear newspapers regularly. About 40 per cent of them read '*Navbharat Times*' and 21 per cent read '*Hindustan Dainik*'. '*Sandhya Times*', an evening newspaper, was read by 6 per cent women under study. This shows great popularity of Hindi newspapers among rural women.

Table 23

Newspapers read or heard by farm women of different lifestyles

Newspapers	Modern (N=74)	Semi-modern (N=192)	Traditional (N=70)	Total (N=336)
Navbharat	42	52	5	40
Hindustan Dainik(Hindi)	29	26	2	21
Sandhya Times (Hindi)	18	4	-	6
Employment News (English)	10	-	-	2
The Hindustan Times(English)	5	-	-	1
Any other	5	-	-	1

Multiple responses.

Table 24

Magazines read or heard by farm women of different lifestyles

Magazines	Modern (N=74)	Semi-Modern (N=192)	Traditional (N=70)	Total (N=336)
Sarita	57	51	-	42
Nandan	36	30	-	25
Grah-Shobha	34	22	-	20
Saptahik Hindusthan	23	24	-	19
Filmi Kaliyan	18	13	-	11
Dharmyug	23	11	-	11
Champak	18	9	-	9
Parag	7	11	-	11
Manorama	8	4	-	6
Mayapuri	2	9	-	6
Soviet Nari	2	2	2	2
Mukta	-	2	-	1
Any other	7	2	-	1

Multiple responses.

Table 24 depicts variety of magazines read by farm women. The women's order of preference for magazine reading was *'Sarita'*. *'Nandan'*, *'Grah-Shobha'* and *'Saptahik Hindustan'*. Table 24 clearly reveals that magazine reading was not a part of traditional women's life style as only 2 per cent out of 70 resorted to magazine reading. Other interesting highlight of the table is the new trend of reading film magazines by farm women. Findings on similar lines were reported by National Readership Survey (1978) and Sarkar (1982). Data show the popularity of *'Nandan'* a magazine essentially for children. Similarly, other children's magazines like *'Champak'* and *'Parag'* were also read by a few farm women.

Largest number of farm women read print media for stories and poems. Other two equally mentioned reasons by farm women were to have 'knowledge about women's life' and 'to get entertainment and happiness'. Similarly, among rest of the reasons given, print media were considered as a source of knowledge. Thus, print media were considered primarily as entertainers and partly as informers. It seems newspapers, magazines and story books will remain vital media in future as well and thus the educational role of print media is bound to increase with the spread of education among farm women. Dhillon (1968), Sanoria and Singh (1976), Kaur (1982) and Saini (1985) in their respective studies also found that reading resulted in gain in knowledge and information. The low involvement of traditional women indicated their inability to understand print media. Some of them, however, gained knowledge through print media which was read out to them by family members or friends. Table 25 gives the details.

Table 25
Reasons given by farm women for using print-media

Reasons	Modern (N=74)	Semi-Modern (N=192)	Traditional (N=70)	Total (N=336)
Read stories and poems	71	57	-	48
Knowledge about women's life	63	48	2	42
Entertains and give happiness	71	45	2	42
Knowledge about current prices	23	41	-	29
Knowledge of politics	42	32	2	28
Knowledgg of city life	52	28	2	28
Knowledge about home crafts	44	28	2	26

(Contd.....)

(Table 25 Contd....)

Reasons	Modern (N=74)	Semi-Modern (N=192)	Traditional (N=70)	Total (N=336)
News about sports	34	22	-	20
Knowledge about agriculture	26	23	5	19
Improves mental level	7	13	2	10
Knowledge of new developments	5	5	-	4
Any other	-	-	-	-

Multiple responses.

The reasons given for not reading print media regularly were lack of free time, bad health, fatigue, 'lack of family support' and 'knowledge not related to life'.

The print media have become popular among farm women and they find them interesting and meaningful. It was, however, realised that newspapers and magazines read by them were essentially urban-oriented and did not cater to their special needs. What is needed now is to publish simple, progressive and relevant magazines and newspapers which could be made available to them at low cost. This will help them to form opinions on important social issues and also make them more progressive.

The study has indicated towards an increasing trend in media utilization among rural women. Although villages around Delhi do not reflect correct picture of the rural situation due to urban influence but, the future is clearly reflected in the trends studied here. With the growing literacy and subsequent exposure to mass-media, the changes in life-style of rural women is bound to creep in. It is a matter of time.

Chapter X

Films--Popular But Controversial Medium

Film is a powerful medium of communication which influences the lifestyle of viewers. The influence of films on social values and modes of behaviour has been a popular field of study in sociology. Easy reach of film to all social strata makes it even more influential medium of communication and social change.

India is the largest producer of various kinds of films in the world. In 1931, Ardashir Irani produced the first Indian film 'Alam Ara'. Till 1981, 15,675 talkie films and 1,268 silent films were produced in 35 languages in India. In the year 1982, 864 cinema films, 123 documentaries and over 100 news-reels were produced, whereas in 1987 about 1000 films (highest in the world) were produced in 20 different languages.

In 1981, there were 10,889 cinema houses in India. Nearly, 54 per cent of the total cinema houses were in Andhra Pradesh, Tamil Nadu, Kerala and Karnataka. About 46 per cent collection of income from sale of cinema tickets came from these states. In 1987, the number of cinema houses in the country had risen to 12,000 with an audience of 700 m/week.

The cinema exercises a tremendous influence on viewers' minds. Maximum films have been made with social themes (83%) followed by crime. The Committee on the Status of Women in India (1974) has reported:

"There is a rapid increase in the emphasis on sex themes in films. The ambivalent attitude of the Indian film producer is illustrated in its projection of the herione who is supposed to embody all the virtues but is usually immodestly dressed and the hero, who is otherwise a good man behaves in a very vulgar fashion on meeting the heroine. The usual Indian film has by and large exploited sex to attract audiences and has degraded the image of the Indian woman".

A survey conducted in Greater Bombay (1957) by the Enquiry Committee on Film Censorship (GOI, 1974) noticed a tendency among adolescents to imitate patterns of behaviour shown in the films. The unhealthy influence noticed were in habits of living and spending, manners and mannerisms including fashions and clothes, hair dressing, speech and behaviour towards the opposite sex and immoral and anti-social practices like stealing, prostitution etc.

Films were found to exercise more influence on youths especially the juvenile and uneducated. A survey conducted by IIMC (1968-69) (GOI, 1974) revealed that 88 per cent of the respondents were film goers with youth in the majority. The study further revealed the categories of films preferred and generally viewed by women as:
(i) films on family life, (ii) musicals and (iii) devotional films. However, girls preferred (i) Western films (ii) musicals, (iii) romantic and tragic films. The respondents had least preference for contemporary films. When asked as to what should be the chief aim and purpose of films in India, 42 per cent of women emphasised the development of social, cultural and religious values in society.

Although several studies found that films help in learning (National Institute of Audio-visual Education, 1961) and Nigam (1964) (GOI, 1974) the over-emphasis on sex and crime has made films a negative factor in socio-cultural change. Therefore, an increasingly large number of films are being censored for indecency and immorality. In 1980, 168 films were certified for showing for adults only whereas in 1978 and 1979 there were 126 and 65 adult films respectively. Indian women regarded treatment of sex and love in films as objectionable and indecent. However, college girls held the opposite view. Men and women agreed that exposure of women in films served only commercial exploitation.

While reviewing the role and influence of Indian films, Committee on Status of Women in India (1974) reported that in its content and treatment of women, films lay more emphasis on sex to draw audience. In most of the social films, woman is invariably assigned a subordinate status in relation to man and thus films continue to perpetuate the traditional notion of women's inferior status. These films have not attempted to educate women about their rights, duties and responsibilities in society.

The recent development of TV industry in India and viewers obvious preference for feature films on TV have made films more popular than before. Now, films are frequently seen in cinema halls and on TV sets. The video boom in the eighties has taken films into the drawing rooms and bedrooms with very high viewership among all classes and age groups.

Besides big private producers of feature films, the National Films Development Corporation is also involved in production and co-production of high quality low-cost films which can be shown on TV as well. During the Seventh Five Year Plan period, Govt. of India intends to spend about Rs. 41.4 crores on production of quality films.

Film Viewing Behaviour of Farm Women

In the study conducted on farm women in Delhi territory, it was found that the farm women were exposed to films through T.V. as well as cinema (theatre). About 40 per cent of the farm women did not see films either on TV or in theatre. Table 26 shows the film viewing behaviour of farm women in detail.

Table 26

Film viewing behaviour of farm women in Delhi villages (%)

Film viewing	Modern (N=74)	Semi-Modern (N=192)	Traditional (N=70)	Total (N=336)
See films in theatre and on T.V.	83	50	46	56
See films on TV only	5	-	-	3
See films in theatre only	3	-	-	1
Do not see films	9	50	54	40

A large number of farm women were found to view films in theatre as well as on TV. Film viewing was more popular among farm women of modern lifestyle as compared to those who had semi-modern or traditional life style.

It was interesting to note about who commonly escorted them **to theatres.** It was found that women never went alone to see films in theatres. About 30 per cent of the movie-going women accompanied their husbands while others went with parents (20%), brothers (17%), sisters (16%), friends (12%), daughters (3%) and sons (2%).

Relevance of Films

During the investigation in Delhi villages, farm women were asked to indicate relevance of films in their life. The intention was to know whether they perceive films as related to life or not. Their responses were divided in three categories viz. 'related to life','partially related to life' and 'not related to life'. The responses have been given in Table 27.

Table 27

Relevance of films to life as perceived by farm women in Delhi villages (%)

Perceived Relevance	Modern (N=74)	Semi-Modern (N=192)	Traditional (N=70)	Total (N=336)
Related to life	60	36	10	36
Partially related to life	35	44	-	32
Not related to life	5	18	85	29
Do not know	-	2	5	3
	100	100	100	100

It can be seen that responses of farm women in three categories were close to each other. Larger number of farm women (36%) felt that films were close to life situations followed by 32 per cent farm women who felt that films were only partially related to life. However, about 29 per cent of them felt otherwise. Maximum number of farm women with modern lifestyle, perhaps, because of their better perception of films, felt the films were closely related to life. This was not the case with farm women belonging to traditional lifestyle. About 85 per cent of them found that films were away from the realities of life. The farm women of semi-modern disposition, however, found enough relevance in films although majority of them found only a partial relevance. It may also be noted that some farm women belonging to semi-modern and traditional life style categories did not know anything about interrelationships between films and life situations.

Impact of Films:

To talk about impact of films is to enter into a very vital area of socio-cultural and economic aspects of life. Films have become the main media of setting fashions and lifestyle. The film makers have exploited the power of films in bringing about desirable or undesirable directed changes in human behaviour. The new wave of consumerism and 'impossible' standards of living including dress, food and cosmetics seem to have their roots in films. In rural areas where films had not reached till recently, the impact was clearly observed. There was a visible impact of outlook of women and their understanding the problems of children etc. Table 28 shows the trends in the women's opinion on impact of films on their life style.

Table 28

Opinions of farm women about impact of films on life style in Delhi villages (%)

Opinion	Modern (N=74)	Semi-Modern (N=192)	Traditional (N=70)	Total (N=336)
Women imitate films and become 'modern'	14	78	92	68
Education for better living	14	75	94	66
Women become fashionable	34	70	80	65
Develop spirit of unity in family ties	14	67	86	61
Help to understand children's problems	14	66	86	60
Inspire better home keeping	70	57	40	56
Film viewing women ignore own cultural norms and values	90	46	26	50

The Delhi study also showed that largest number of farm women believed that women imitate films and thus change in outlook towards modernity takes place. This opinion was shared more by the semi-modern and traditional women who held films squarely responsible for all that is happening to the lifestyle of farm women. The educational aspects of films were widely appreciated by the farm women who also shared the opinion that film viewers develop spirit of unity in family ties. The modern farm women attached more importance to impact of films on home keeping and fashion. It was, however, surprising to note a large number of modern and semi-modern women felt that films were responsible for eroding norms and values in society. And yet these were the two groups of women who comparatively saw more films. This is, perhaps, an example of magic spell the films cast on viewers. May be film viewing is an addiction.

It can, therefore, be concluded that traditional women felt strongly about the impact of films on the lives of the people. Impact of films was considered positive as well as negative. Farm women did mention that much depended on the choice of right kind of films. They opined that films with sex, love and violence certainly had negative influence on lifestyle of farm women.

Religion or Romance?

What type of films farm women preferred to see? There were seven different types of films which could be indicated to the women of Delhi territory. They were asked to state types of films they would like to see. Their favourites varied from religious to romantic films and of course, others like socio-family drama, national integration, social new wave, historical and documentry films. The following table shows the details.

Table 29
Preferential film viewing pattern of farm women in Delhi villages (%)

Type of films	Modern (N=74)	Semi-Modern (N=192)	Traditional (N=70)	Total (N=336)
Religious	67	72	54	67
Socio-family drama	67	54	20	50
National integration	43	46	11	38
Social new wave	51	24	8	26
Historical	21	27	5	21
Romantic	46	11	-	16
Documentry	10	10	5	9

Multiple Responses.

It is clear from Table 29 that religious films were most popular among farm women followed by films on socio-family drama, national integration and social new wave. This means new wave films were liked by farm women as much as by their urban counterparts. However, romantic and documentary films were found to be least popular types. The findings revealed that farm women by and large were religious minded and liked to see epic stories in films. Socio-family drama were popular because of identical 'film-family' situations. Romantic movies, although not popular, were comparatively liked more by modern farm women. The findings showed that the popularity of religious and socio-family drama could really be exploited to achieve desirable results. The documentary films, though less popular, if prepared to convey messages at the mental level of farm women will have better impact.

Thus, the farm women had mixed opinions about films. About one third of them said that films were related to their life. Another one third thought that they were partially related to life whereas rest of the one third thought that they were not related to life. Similar findings were reported by IIMC. (1969) (GOI, 1974) about sex and love in films. It was pointed out that one third women regarded sex and love in films objectionable and another one third regarded this as normal in films. It can, therefore, be concluded that films were accepted as completely relevant by one third of the farm women while the rest of them felt that films were away from realities of life. More of modern farm women positively empathised with various themes of the films. A survey conducted in Bombay city (1957) (GOI,1974) had reported similar findings on impact of films on viewers.

Undoubtedly, film was the most controversial medium of communication among the farm women in Delhi villages. Both positive and negative influences were reported by farm women.

Chapter XI

What Do Farm Women Think About Social Issues?

Mass media are not substitutes for interpersonal communication which is more dominant in rural societies specially among women. Their communication with elderly women at home, neighbours, friends, relatives, etc. is more crucial than the use of mass media. Nevertheless, the use of mass media strengthens the effectiveness of interpersonal communication. A stage has come where communication media keep shaping the lifestyle of farm women and influence their thinking on social issues of consequence.

Before discussing what farm women think about the social issues, and how these are related to their lifestyle, let us mention about what these issues are and how they are important to farm women.

The issues are mainly related to marriage, inheritence of property, socio-economic and political status of women and their status of health. These issues reflect in totality the woman's status as it exists in the rural society. The opinions farm women hold on these issues are reflections of change in their thinking which is perhaps steadily creeping in. Most of the women's upliftment programmes either aim at educating women or developing them on these lines.

The opinions of women on these issues also reflect the quality of socialization of farm women and are ultimately responsible for their achievements, social harmony and frustrations. The issues have assumed importance even in the legal context in which several laws have been enacted to protect the rights of women as related to these issues.

The women respondents in the study conducted in Delhi villages readily expressed their opinions on the various socio-economic and political issues put before them. Their responses duly categorised according to lifestyle have been explained under each issue given in the following pages.

Social Issues

Although most of the issues included have bearing on social life of farm women, some issues were especially selected as indicators of varied opinions expressed by farm women belonging to different life style groups.

Co-education: The opinion of farm women about co-education was divided wherein half of them favoured co-education. Most of the traditional farm women (97%) expressed their opinion against co-education while 37 per cent of semi-modern and 13 per cent modern farm women opposed it. Those who disfavoured it believed that co-education leads to free mixing of opposite sexes at an immature age which creates problems. Those farm women who favoured co-education system mostly belonged to modern lifestyle and therefore, felt that they could study together like brothers and sisters which may lead to better understanding. The Committee on Status of Women in India (1974) also mentioned that the mixing of opposite sexes at primary and middle school level was acceptable to the officials dealing with education in rural areas while it was not so at post middle level of education.

Legal Status

Indian Independence brought complete legal equality for women in India. One of the main characteristics of modern society is a heavy reliance on law to bring about socio-economic changes, particularly in those countries which had for centuries been under foreign rule and attained Independence after a long struggle.

The Indian Constitution guarantees equality of opportunity and rights to both sexes against discrimination. It also recognises the inequality suffered by women and contemplates that it might be necessary to make special provisions in their favour. Yet, a lot needs to be done. Education is the main instrument of change. While spread of formal education shall take time, mass media can be used for educating rural women even in remote corners of the country. The media are popularising positive opinions on legal, social and political issues among the rural masses.

Economic Issues

Right to Property: Whether women should ask share in parental property or not is a sentimental issue for Indian women. Many women for the fear of loosing contacts with the parental family avoid to subscribe to the idea of asking share in the parental property.

According to the study in Delhi villages, about 63 per cent farm women disfavoured the idea of asking share in parental property. While 22 per cent favoured it, 15 per cent did not express any opinion. About half of the modern farm women under study favoured this idea as against 20 per cent semi-modern and 3 per cent traditional farm women. The main reason for disfavouring the idea was that it was against social norms. Those women who subscribed to this view thought they will loose respect in the brothers' family if they ask for a share in parental property. It was thought to be already given in the form of dowry. Moreover, she would get the property which her husband gets as his share. Only when there was no brother, the woman was expected to claim her share. On further probing, it was found that about 25 per cent farm women preferred that only brothers should get the parental property. However 41 per cent farm women opined that daughters were like sons so they should get equal share. A small number felt that daughters need to be given education and not property.

Equality of wages: It has been widely accepted that men and women have been gifted with equal physical and mental powers. Therefore, several legal measures have been taken to ensure equality of wages between men and women.

Majority of farm women (82%) in Delhi villages favoured the idea of men and women getting equal wages, 8 per cent disfavoured and 10 per cent had no idea about it. More of modern farm women (97%) favoured the idea as compared to semi-modern (92%) and traditional (40%) farm women. It shows farm women were aware of equal rights of women as far as wages were concerned.

Saving money for old age: The issue of economic freedom of farm women has been a subject of discussion amongst sociologists and economists alike. Women respondents in villages under study freely expressed their views in favour of economic freedom for women. Majority of farm women (82%) favoured women opening their *separate bank accounts*. Those who disfavoured this idea were mainly traditional (74%). More than half of the farm women favoured the idea of saving for old age. This was perhaps because the joint family system was gradually on its way out even in the rural areas. Those who opposed this view mostly belonged to lower education group and those who did not preceive the changing pattern of family system in the rural areas.

Loans for Social and Religious ceremonies: It is a common practice to see farm families taking loans on social and religious ceremonies. It was encouraging to find that 60 per cent of farm women disfavoured the idea as against 28 per cent who favoured it and 12 per cent had no opinion on it. It was more of modern farm women (73%) who disfavoured it as compared to 70 per cent semi-tradtitional and 20 per cent traditional farm women. Those who disfavoured this felt it was the basis of all troubles since a farmer who was generally poor became poorer by taking unproductive loans.

Marriage Related Issues

Widow remarriage: In the study conducted in Delhi villages, about 60 per cent farm women favoured widow remarriage. About 90 per cent modern, 62 per cent semi-modern and 13 per cent traditional women favoured the idea. Many preferred so because they thought remarriage will give the widow a chance to resettle in life and give her encouragement. It was further found that a few traditional women who did favour the idea of widow remarriage wanted the marriage within the family (husband's brother) to save family honour. However, modern and semi-modern women, did not stipulate any such conditions for widow remarriage. A small number of women felt that it was the widow's choice to marry or not to marry and others need not force her. A few farm women believed that widow remarriage was difficult because of dowry problem.

Fifty-nine per cent of farm women favoured the idea of widows wearing good and neat clothes. However, more of traditional women wanted them to lead a simple and pious life as they believed women should dress up only to please husbands.

Polygamy: In Delhi study, acceptance of second marriage of the husband was a crucial social issue which confronts farm women. On the whole, 75 per cent of them disfavoured this idea and about 17 per cent had no opinion to offer on this issue. As against the prevailing tradition in the past this practice is now becoming extinct. The Hindu Marriage Act (1955) lays down the principle of monogamy for Hindus thus putting legal binding on 88 percent of the Indian population. However, Muslim law in this regard has recently raised some controversy on this important issue.(GOI, 1974)

The Committee on Status of Women in India (1974) conducted a survey and found that 85 per cent of men and 96 per cent of women are in favour of compulsory monogamy. Thus, the Committee favoured monogamy as a rule for all communities in India. Yet, due to lack of education and kndowledge, many Hindu women suffer due to bigamous marriages. The Delhi study found all traditional farm women opposing bigamy except under exceptional circumstances such as when first wife could not bear any children or when the younger brother died and his wife was to be married to elder brother to preserve family honour.

Divorce : Divorce in India is not a common practice. However, percentage of divorced and separated females is 0.33 per cent, (census 1971, Govt. of India, 1974). A 1961 Census Survey in 587 selected villages showed that the incidence of divorce was higher in village communities than urban communities. It was highest amongst Muslims (6.06%), followed by Hindus (3.21%), Buddhists (3.07%), Jains (1.68%), Sikhs (0.91%) and Christians (0.41%). The main causes for divorce were found to be adultery and barrenness. Extreme poverty was also found to be a cause for divorce.

The rights of divorced women have been protected by some legal measures viz. Hindu Women's Right to Residence and Maintenance Act (1946), Hindu Marriage Act (1955), etc. (GOI, 1974)

The opinion of farm women on divorce in Delhi villages was divided. Almost half of them favoured the idea of Women's right to divorce whereas 38 per cent disfavoured and 10 per cent had no opinion on this issue. Majority of modern farm women (81%) favoured this idea as against 56 per cent semi-modern and 8 per cent traditional farm women. Those who disfavoured the idea did so as they believed that a woman should adjust with her in-laws under all circumstances and never entertain the idea of divorce. Even those who favoured divorce felt that it should be resorted to under extreme circumstances.

Although the idea of free intrmixing of sexes is not encouraged, in Delhi villages only 48 per cent of the farm women disfavoured *girls choosing their own life partners.* The idea was, however, favoured by 92 per cent modern, 50 per cent semi-modern and none of the traditional farm women. Actually, it was considered against their culture and girls having such ideas were looked down upon.

Issues Related to Family Health

Some of the important health practices followed in rural areas have been explained here.

Family planning: The knowledge of family planning methods help women to achieve a better life for themselves and their family. However, the opinion of farm women about family planning is an important precursor of use of family planning methods. Studies have indicated that most of the women who adopt family planning methods were motivated by small family norm.

In Delhi villages, it was found in the pilot study that most of the women did not have much reservation against simple family planning methods and their usage. However, they had varied opinions about abortion. About 44 per cent farm women did not favour the idea of abortion as against 21 per cent who favoured and the rest refrained from giving any opinion mainly because of ignorance. Again it was more of traditional farm women (88%) who disliked the idea most. The reasons given varied from abortion being bad for health (27%), it was undesirable in the society (12%), other family planning methods are better (8%) and that it was a sin (11%). About 5 per cent farm women felt that those women who go for abortion do not enjoy good reputation. Yet about 17 per cent farm women felt that it was better than having unwanted children. On the whole, those who favoured abortion as a practise to terminate unwanted pregnancies belonged to modern lifestyle category.

Feeding and Childcare Practices

Mass media often give coverage to feeding and child care practices. Farm women were asked to give their opinions on them. Most of the farm women (77%) were convinced that a woman should *rest for 40 days* after child birth to recoup normal health, 18 per cent had no opinion on it and only 5 per cent thought this rest was not required. All traditional farm women favoured minimum forty days rest as against 74 per cent semi-modern and 65 per cent modern farm women.

Majority of farm women (68%) favoured *breast-feeding* whereas 25 per cent had no opinion and 7 per cent disfavoured it. Number of modern farm women (73%) who favoured breast-feeding was more than semi-modern (59%) and traditional farm women (32%). This shows that the coverage by mass media on encouragement of breast-feeding was having a desired impact on modern farm women. A fairly large number of farm women (62%) felt that pure ghee should be fed to mothers after child birth to have better lactation.

An important issue in child nutrition is the correct stage to start *supplementary feeding*. It is normally recommended at 5 months of age. More than half of the farm women (55%) favoured starting supplementary feeding at the age of five months whereas 33 per cent did not favour it and 12 per cent had no opinion on it. It was revealed that more of modern farm women (84%) favoured the idea as compared to 59 per cent, semi-modern and traditional (11%) farm women. This shows that mass media were effective in convincing modern women about modern childcare practices.

Issues Related to Domestic Chores

Traditionally women share larger part of the household jobs in rural families. This pattern is more significantly noticed in those families which belong to lower socio-economic status. Moreover, those household works which involve drudgery and hard work are given to women. Even women are made to believe of their lower position in the family and that only they are required to do all the work.

In Delhi villages, about 44 per cent of the farm women even disfavoured men coming to kitchen to help the women. About 21 per cent of the farm women expected men to help women in the kitchen. Surprisingly, 34 per cent of the farm women did not express any opinion on this issue. Those who favoured men's working in the kitchen expected it only in emergency e.g. when the wife was sick or there was too much of pressure of work at home. The variation amongst the farm women belonging to different categories of lifestyle are not very significant although a trend towards more modern farm women expecting men to help them in the kitchen is clearly visible.

Even amongst the women in a household, work distribution is sometimes not fair. Study in Delhi villages showed that women were equally divided when it came to the question of whether *daughter-in-law was supposed to do all the household* work. Out of the total (336) women, 51 per cent favoured it. Those who disfavoured it felt that household chores should be more or less equally divided among women in family. It was interesting to note that all the traditional farm women favoured daughter-in-laws doing all the household work, whereas all modern farm women disfavoured this idea. Since modern farm women happened to be of younger age group and in the category of daughter-in-law themselves, one can imagine the amount of clash which can take place because of such opinions. This can even lead to split of the joint families.

The *use of modern gadgets* in homes not only makes life comfortable but also gives status to a household. More and more domestic gadgets are reaching rural homes such as stove, pressure cooker, electric press and others. They reduce drudgery of rural women and make their life a little worthwhile. In Delhi villages, a large number of farm women (67%) favoured the use of such gadgets, 14 per cent disfavoured and 19 per cent had no opinion. Surprisingly, largest number of traditional women (83%) favoured them as against 68 per cent modern women.

Voting in election should necessarily be an individual decision. Almost all farm women (98.8%) irrespective of their life style favoured women casting votes to select a government. Almost all were in favour of having women's leaders to fight for the cause of women.

Appendix I
Measurement of Life Style

Instructions

Here is a set of 46 statements which you can use to assess quality of lifestyle of farm women. The woman respondent be asked to read each statement carefuly and give response in the form of a check mark (/) on any one of the five points indicated against each statement. The statements carry weightages in bracjets which show whether the statement indicates modern outlook of semi modern or traditional outlook of the woman respondent. Weightages less than 2.5 indicate trend towards modernity (+ positive) and more than 2.5 indicate trend towards traditionality (- negative).

Scoring:- After each statement has been marked as per instructions given above, the researcher has to calculate the lifestyle.

Now follow the following steps:-
1. Multiply the weightage of the statement in the bracket by the numerical indicated on the fice point continuum.
2. The continuum shows two kinds of numericals. In case the statement has values less than 2.5, the numericals will be as follows:-

Strongly agree	:	1
Agree	:	2
Undivided	:	3
Disagree	:	4
Strongly disagree	:	5

In case the statement has a weightage of over 2.5, the numerical with which multiplication has to be done will be as follows:

Strongly agree	: 5
Agree	: 4
Undivided	: 3
Disagree	: 2
Strongly disagree	: I

3. Sum up the products of multiplications of 46 statements.

4. Since the number of statements is 46, please divide the final product by 46. This is the lifestyle score of the respondent.

5. This score will range between 2.2 to I0.5. The categories of lifestyle will be as follows:

   (i) Modern lifestyle - score ranging between 2.2 to 5.0
   (ii) Semi-modern lifestyle - score ranging between 5.I to 8.0
   (iii) Traditional lifeestyle - score ranging between 8 to 0.5.

Quality of Life

	Strongly Agree (+1&-5)	Agree (+2&-4)	Undecided (+3 & -3)	Disagree (+4&-2)	Strongly Disagree (+5&-1)

1. I have somewhat old fashioned tastes and habits (2.8)

2. Religion is major part of my life (3.4)

3. My days seem to follow a definite routine (2.5)

4. I feel my children are getting better education than me (1.8)

Work and Activities

5. I would get more involved in my community if I had time (1.8)

APPENDIX

	Strongly Agree (+1&-5)	Agree (+2&-4)	Undecided (+3 & -3)	Disagree (+4&-2)	Strongly Disagree (+5&-1)

6. I like to try out new recipes (1.5)

7. I am the kind of person who will try almost anything once (1.3)

8. I feel like learning new things in leisure (1.6)

9. I like to wash clothes near pond since there I happen to meet my friends (3.0)

10. I like to work in the fields (3.0)

11. I take great pride in the appearance of my house (1.5)

12. Seeing present circumstances, I work on the field as much as the men of my home (2.0)

13. I feel I get more leisure time than the times of my mother-in-law (2.0)

14. I am very fond of stitching and embroidery (2.1)

15. I do not enjoy sweeping the house (1.4)

16. I feel only men of the house have the previlege of spending money (3.2)

	Strongly Agree (+1&-5)	Agree (+2&-4)	Undecided (+3 & -3)	Disagree (+4&-2)	Strongly Disagree (+5&-1)

17. I think women should do village welfare activities in their spare time (1.3)

Media Use

18. Viewing of television is a waste of time (2.9)

19. Reading book is hard work (2.5)

20. I like romantic movies (1.2)

21. I cannot understand why anyone would go out of village for entertainment, when such facilities are available in the village itself (2.5)

22. Where is the time with me to go out for entertainment? (2.6)

23. I want to see only those films which are related to my life (2.3)

24. Listening to radio is an essential part of my life (1.3)

25. It is better for me and my family not to see films because one learns only bad things from it (3.1)

APPENDIX

	Strongly Agree (+1&-5)	Agree (+2&-4)	Undecided (+3 & -3)	Disagree (+4&-2)	Strongly Disagree (+5&-1)

Community Identification

26. I prefer a community where people mind their own business (1.7)

27. I would like to spend my entire life in the village (3.4)

28. Men and women should work together for the upliftment of the villages (1.6)

29. I feel only elderly women should participate in social and religious ceremonies of village (2.8)

Social Attitude

30. A girl should learn all household work before marriage (2.8)

31. I prefer a traditional marriage with the husband providing for the family and the wife running the house and taking care of children (3.1)

32. A woman's place is at home only, she should not take up job outside (3.4)

33. A home is incomplete without a son (3.4)

	Strongly Agree (+1 & -5)	Agree (+2 & -4)	Undecided (+3 & -3)	Disagree (+4 & -2)	Strongly Disagree (+5 & -1)

34. I feel two children are enough in the present times (1.4)

35. I feel dowry system should be discouraged in villages (1.4)

36. I will have no objection if my children go for inter-caste marriage (1.0)

37. Like olden days, I do not favour early marriage for my children (1.3)

38. I still consider 'harijans' as untouchables (3.4)

39. I consider 'Purdah' system an outdated custom (1.3)

40. I do not like to stay in a joint family (1.1)

41. I favour higher education for my daughters (1.2)

Family Situation

42. I often feel I live for my children (2.4)

43. Every woman should live as per her husband's wishes (3.1)

44. Husband and wife should

APPENDIX

| | Strongly Agree (+1&-5) | Agree (+2&-4) | Undecided (+3 & -3) | Disagree (+4&-2) | Strongly Disagree (+5&-1) |

take joint decisions related to family matters (1.4)

45. It does not appear good if mother-in-law does household work when daughter-in-law is there (3.0)

46. I feel in our home boys are preferred over girls (2.7)

BIBLIOGRAPHY

Anon. (1978) All India National Readership Survey. (Delhi City) Indian Market Research Bureau. IIMC, New Delhi.

Anon. (1984) An Analysis of the Situation of Children in India. U.N. Children's Fund. Regional Office for South Central Asia, New Delhi.

Anon. (1975) Audience Research Profile, TV Centre, Srinagar.

Anon. (1985) Comparative analysis of Male and Female Enrolment and Illiteracy, UNESCO.

Anon. (1987) Competition Success Review Year Book.

Anon. (1987) Critical Issues on the Status of Women - Suggested Priorities for Action. Advisory Committee on Women's Studies, ICSSR.

Anon. (1985) Economic Intelligence Service Basic Statistics relating to the Indian Economy. Vol. 1, All India Centre for Monitoring Indian Economy, Bombay.

Anon. (1982) Economic Situation and Prospects of India, World Bank, Washington, D.C.

Anon. (1985) Education of Girls in Asia and the Pacific Reporting of Regional Review Meeting. UNESCO.

Anon. (1982) Family Welfare Programme in India, Year Book, Ministry of Health and Family Welfare, Govt. of India.

Anon. (1979) Farm School on the AIR, Audience Research Unit, All India Radio, Delhi.

Anon. (1988) Manorama Year Book.

Anon. (1985) Education of Girls in Asia and the Pacific Report of Regional Review Meeting. UNESCO.

Anon. (1983) Media and Women, Mainstream, 22(15)

Anon. (1984) Press in India, Press Registrar of India, New Delhi.

Anon. (1973) Profile and Use of Leisure Report on Sample Survey Among Needs of TV Household in Delhi City. Audience Research Unit, TV Centre, AIR.

BIBLIOGRAPHY

Anon. (1967) Profile of TV Listeners, Audience Research Unit, TV Centre, AIR, New Delhi.

Anon. (1978) Report on the 'arm School on the AIR' Programme of AIR, Tamilnadu Agricultural University, Coimbatore.

Anon. (1975) Report on the Farm School on the AIR, Programme of AIR, Vijaywada Audience Research Unit, AIR, Hyderabad.

Anon. (1976) A Sample Survey Report on Profile of Prospective TV Audience Research Unit, AIR, Madras.

Anon. (1978) A Sample Survey Report on Profile of Perspective TV Audience in Madras. Audience Research Unit, AIR, Madras.

Anon. (1985) Sample Registration System, Vital Statistics Division, Office of the Registrar General, Ministry of Home Affairs, Govt. of India.

Anon. (1983) Science and Technologies for Women. Complied by Centre of Science for Villages, Wardha, Sponsored by Department of Science and Technology, New Delhi.

Anon. (1984) Statistical Abstract: India - New Series. Central Statistical Organization, Department of Statistics, Ministry of Planning, Govt. of India.

Anon. (1984) Statistical Digest for Asia and the Pacific, Office of the Regional Coordinator for UNESCO Programme in Asia and the Pacific, Bangkok.

Anon. (1986) Statistical Year Book, UNESCO.

Anon. (1980) Survey of Causes of Death, Office of the Registrar General, Ministry of Home Affairs, New Delhi.

Anon. (1979) Survey on Infant and Child Mortality. Office of the Registrar General, Ministry of Home Affairs, New Delhi.

Anon. (1985) Towards Equality of Educational Opportunity UNESCO. Regional Office for Education in Asia and the Pacific Bangkok.

Anon. (1974) Towards Equality. Report of the Committee on Status of Women in India. Department of Social Welfare, Ministry of Education and Social Welfare, Govt. of India.

Anon. (1980) Vital Statistics of India, Office of the Registrar General, India, Ministry of Home Affairs, New Delhi.

Anon. (1985) Women and Development, Papers presented in The International Seminar on Women and Development. Society for International Development, Rajasthan Chapter, Jaipur.

Anon. (1977) Women and Media in Asia, Centre for Communication Studies, The Chinese University of Hong Kong, Hong Kong.

Aggarwal, V. (1972) A Study of the Food Habits of People in a Selected

Rural Community: Studies of Rural Community. Lady Irwin College, New Delhi.

Annamalai (1979), A Study on Utilisation of Farm Information Sources in the Adoption Process, M.Sc. Thesis, T.N.A.U., Coimbatore.

Ambastha, C.K. and Singh, K.N. (1975) Communication Pattern - a Systematic Analysis. Indian Journal of Extension Education, 1&2.

Badrinarayanan, P.A. (1977) A Study of Farm Broadcast Listening Behaviour of Small Farmers, M.Sc. Thesis, Tamilnadu.

Bhandari, S (1972) Reactions of Rural Youth Towards 'Yuv-vani', Studies of Rural Community, Lady Irwin College, New Delhi.

Bhani Ram (1981) A Critical Analysis of Farm School on the AIR an Instructional Programme of AIR, New Delhi, Ph.D. Thesis, I.A.R.I., New Delhi.

Bhaskaran, C. (1976) A Study of Socio-metric Identification of Opinion Leaders and Their Characteristics in a Progressive and Non-prograssive Village in Kanya Kumari District of Tamil Nadu, M.Sc. (Ag) Thesis, Department of Agrl. Extension, University of Agricultural Sciences, Bangalore.

Chahil, R. (1972) A Comparative Study of Effectiveness of Radio, Television and Pamphlets in the Communication of Agricultural and Family Planning Information. Studies of Rural Community, Lady Irwin College, New Delhi.

Chandrakandan, K. (1981) Study on the Impact of Farm School on the AIR on Scientific Farming - A Project Study. Department of Agricultural Extension and Rural Sociology, Centre for Agricultural and Rural Development Studies, Tamil Nadu Agricultural University, Coimbatore.

Chatterjee, R.K. (1973) Mass Communication, National Book Trust, India, New Delhi.

Chabra, Rani (1983) Women and the Media, Indian Express (Jan. 28th).

Chowla, N.L. (1983) Change: An Indian Overview. Media Asia, Vol. 10(2)

Daswani, T.C. (1984) Women in Media and Attitudes, Vidura (Oct.), 21(5).

Dave, S.C. (1975) The Radio Listening Habits of Farmers of Anand Taluka of Kaira District Hearing Rural Radio Programmes Broadcast by AIR Ahmedabad Baroda Station, M.Sc. Thesis, M.S. University, Baroda.

Dey, P.N. (1968) Relative Effectiveness of Radio and Television on Mass Communication Media in Dissemination of Agricultural

Information, M.Sc. Thesis, I.A.R.I., New Delhi.

Dhadhal, B.R. (1973) Impact of Rural Radio Programme Broadcast by AIR, Rajkot in Disseminating Agricultural Information to the Farmers in Junagadh District of Gujarat State, M.Sc. Thesis, G.A.U., Gujrat.

Dhaliwal, A.J.S. and Sohal, T.S. (1967) Preferences of Radio Programmes by Rural Listeners Indian Journal of Extension Education, Vol. 3(3)

Dhillon, J.S. (1968) Impact of Popular Periodicals of Punjab Agricultural University, Ludhiana on the Dissemination of Improved Agricultural Practises to Farmers of Punjab State, M.Sc. Thesis, PAU, Ludhiana.

Dhillon, D.S. (1978) An Appraisal of the Correspondence Course Programmes for the Farmers (general) run by the Punjab Agricultural University in Punjab State, M.Sc. Thesis, PAU, Ludhiana.

Edeani, D.O. (1980) Critical Predictors of Orientation to Change in a Developed Society. Journalism Quarterly, (Spring).

Garg, N. (1972) An Investigation into the Role Expectations and Role performances of Key Female Members in Selected Agricultural Families, Studies of the Rural Community, Lady Irwin College, New Delhi.

Gokhale, L.M. (1984) Towards another Development with Women and Mass Media Experience, Vidura (Oct.)) 21(5).

Inkles, A. and Smith, D.H. (1974) Becoming Modern: Individual Change in Six Developing Countries, Cambridge, Harvard University Press.

Hassan, A.M. (1983) The Mass Media as an Agent of Change in Malaysia, Media Asia, Vol. 10(2).

Jalihal, K.A. and Srinivasmurthy, J. (1974) Some Aspects of Evaluation of Farm Radio Programmes in Karnataka, Ph.D. Thesis, UAS, Bangalore.

Jha, R. (1978) Study of Delhi TV: Retrospect and Prospect, M.Sc. Thesis, Division of Agricultural Extension, I.A.R.I., New Delhi.

Joanne Leslie (1977), Five Nutrition Projects That Use Mass Media Development Communication, Report No. 20.

John Knight, A. (1973) A Study of the Relative Effectiveness of Three Modes of Presentation Preferences, Listening and Post Listening Behaviour of Farm Broadcast Listening, Ph.D. Thesis, I.A.R.I., New Delhi.

Kakar, R. (1972) A Comparative Study of the Nutritional Status of Urban and Rural Preschool Children with Special Reference to

Weaning Practices. Studies of Rural Community, Lady Irwin College, New Delhi.

Kamath, M.G. (1973) TV and Social Change - role of TV in Agriculture, Communication, (3)

Kaur, R. (1970) Impact of Television on Farm Women, M.Sc. Thesis, I.A.R.I., New Delhi.

Kaur, S (1982) Comprehension and Use of Information - a Study of Correspondence Course for Farm Ladies in Punjab, Ph.D. Thesis, PAU Ludhiana, (Pb.)

Khajapeer (1978) A Study of academic performance of Farmers Functional Literacy Programme participation in relation to some Sociopsychological Factors, Indian Journal of Adult Education 11 (1&2).

Khandekar, P.R., and Mathur, P.N. (1975) Effectiveness of 'Unnat Krishi', Farm Magazine as related to different Categories of Readers, Indian Journal of Extension Education 11. (1 & 2)

Khattar, J.K. (1972) A Socio Economic Study of the Impact of Green Revolution on the Expenditure Pattern, Studies of Rural Community, Lady Irwin College, New Delhi.

Kinyanjui, P.E. (1972) Recent Developments in Radio Correspondence Education in Kenya, Convergence, 5 (2).

Kishore, D. (1968) A Study of Effectiveness of Radio as a Mass Communication Medium in Dissemination of Agricultural Information, Ph.D. Thesis, I.A.R.I., New Delhi - 110012.

Koshy, S. and Bhagat, R. (1980), Some Aspects of Nutrition Education Associated with Feeding Infants and Toddlers in Rural Farm Families, Indian Journal of Adult Education, 41(3).

Koshy, S. and Sarin, R. (1972) Values Influencing Rural Families in Their Food Procurement, Indian Journal of Extension Education, Vol. 6(2).

Kumari, Abhilasha (1981) Adverse Media Portrayal of Women will Change only Social Consciousness of Women Changes, Communication and Culture (Sept-Oct).

Kuthiala, B.K. (1981) Women and Mass Media: Their Media Participation, Media Projection and Media Consumption. Communication and Culture (Sept-Oct).

Lerner, D. (1958) The Passing of Traditional Society: Modernising the Middle East. New York Free Press.

Lerner, D (1963) Towards a Communication Theory of Modernisation. In Lucien Wrye, ed. Communications and Political Development, Princeton University Press. Princeton, N.J.

BIBLIOGRAPHY

Mcnelly, J.T. (1970) Communication in the Development Process. International Communication, New York: Hastings House Publishers.

Marty Chen et.al (1987) Indian Women - A Study of Their Role in Dairy Movement. Shakti Books, Vikas Publishing House Pvt. Ltd, New Delhi.

Malik, Amita (1981) Women on the Media, Indian Express (Dec. 13th).

Majumdar, Modhumita (1987) Women and Media, Mainstream Annual, 20 (1-5).

Masani, M. (1976) Broadcasting and the People. National Book Trust, India, New Delhi.

Mathew, A. (1982) How Effecitve is the Radio in Educating Rural Mothers? The Indian Journal of Home Science, 14(3).

Mehta, N. and Khanna, K. (1972) Responses of Rural Women Towards 'Grameen Mahilon Ka Karyakram', a Radio Broadcast. Studies of the Rural Community, Lady Irwin College, New Delhi.

Mishra, A.N. (1967) Impact of Television on Farmers, M.Sc. Thesis, IARI, New Delhi.

Mishra, S.P. (1979) A Study of Farm Entrepreneurship in a Backward District of Bihar, Ph.D. Thesis, IARI, New Delhi.

Muis, A. (1983) Some Implications of Television Exposure among Traditional Peasants: A Case from South Sulawesi Island. Media Asia, 10(2).

Murthy, A.S. (1969) Social Psychological Correlates in Predicting Communication Behaviour of Farmers, Ph.D. Thesis, IARI, New Delhi.

Singh, S.N. (1969) A Study on Adodption of High Yielding Varieties and Investment Pattern of Additional Income by the Farmers of Delhi Territory. Ph.D. Thesis, IARI, New Delhi.

Nigam, S.N. (1964) An Experimental Study of the Effectiveness of Instructional Films, M.Ed. Dissertation, Central Institute of Education, Delhi University, Delhi.

Padmanabha, P. (1982) Mortality in India: A Note on Trends and Implications. Economic and Political Weekly, Bombay.

Padmanabha, P. (1971, 1981) Census of India Registrar General and Census Commissioner, India.

Patel, C.S. (1976) Functioning of Khedut Charchamandal organised by Farmers Training Centres, Thasra District Kaira Gujarat State, M.Sc. Thesis, College of Agriculture, Gujarat Agriculture University, Ahmedabad.

Proceedings and Recommendations of All India Agricultural Information Communication Workshop, 10th-13th August (1987), Banaras Hindu University, Varanasi (U.P.) Sponsored by Directorate of Extension (Farm Information Unit) Department of Agriculture and Cooperation, Ministry of Agriculture, New Delhi.

Puri, K. (1972) Rural Listeners Reactions Towards the Radio Rural Broadcast of the Delhi 'A' Station of All India Radio - A Survey. Studies of Rural Community, Lady Irwin College, New Delhi.

Radhika (1981) Women and Mass Media of Kerala. Communication and Culture (Sept-Oct) 2/3.

Rajamani, M. (1981) Impact of Farm Broadcast on Two organised Groups of Listeners: A Comparative Analysis, M.Sc. Thesis, IARI, New Delhi.

Ramson, Mark (1977) Three Media Strategies used in Nutrition Education, Development Communication Report No. 20.

Rao, B. N. (1984) Development of Salt Fortification Programme to Prevent Iron Deficiency in India, National Institute of Nutrition, Hyderabad.

Reddy, S.R. (1977) Attitude of Farmers Towards Farm Radio Programmes, M.Sc. Thesis, A.P.A.U., Hyderabad.

Rogers, E.M. and Svenning, L. (1969) Modernization among Peasants - The Impact of Communication, Holt Rinehart and Winston, Inc., New York.

Sadamate, V.V. (1975) Krishi Darshan Viewing Behaviour of Farmers and International Comparison of Television with Other Sources of Farm Information, M.Sc. Thesis, I.A.R.I., New Delhi.

Saini, G.S. (1970) Impact of Agricultural Broadcasts from All India Radio Jullundur on Rural Listeners, M.Sc. Thesis, PAU Ludhiana.

Sakaya, S.K. (1973) A Study of the Farm Radio Listening Characteristics of Radio Owning Young and Adult Farmers in Nepal. M.Sc. Thesis, IARI, New Delhi.

Sanoria, Y.C. and Singh, K.N. (1976) Communication Sources and Their Credibility, Indian Journal of Extension Education 13 (3&4).

Sandhu, A.S. (1970) Characteristics, Listening Behaviour and Programme Preferences of the Radio Owning Farmers in Punjab, Ph.D. Thesis, PAU, Ludhiana.

Sarkar, K. (1982) Women Journals in India. Mass Media in India. Research and Reference Division, Ministry of Information and Broadcasting, Govt. of India.

Schneider, A. et.al (1970) Communication use in decision on rural credit in South Brazil, Journalism Quarterly, 47.

Schwartz, S.H. (1981) A General Psychographic Analysis of Newspaper use and Life Style. Journalism Quarterly (Spring).

Sekhon, I. (1970, 1972) The Effectiveness of Television as a Medium of Communication for Importing Scientific Know-how to the Farmers. Indian Journal of Extension Education, 6(1&2).

Shahi, P.N. (1977) A Study on Audience Profile of Satellite Instructional Television Experiment (SITE) and its Technocultural impact in Gandak Command Area of North Bihar. Ph.D. Thesis, IARI, New Delhi.

Shankariah, Ch. (1969) A Study of Differential Communication Patterns in a Progressive and Non-Progressive Village, Ph.D. Thesis. I.A.R.I., New Delhi.

Shivpuri, D.S. (1972) A Study of Content of Farm Coverage by the Punjab Press, M.Sc. Thesis, PAU, Ludhiana.

Shrivastava, K.M. (1981) Women in Media Talk. Communication and Culture (Sept-Oct) 2/3.

Singh, J. (1971) A Study of the Factors Influencing the Viewing Behaviour of the Farmers towards Agricultural Programmes on Televisions in Delhi Villages, Ph.D. Thesis, I.A.R.I., New Delhi.

Sinha, B.P. (1970) A Study of Some Motivational Factors in Diffusion of Farm Information through Television, Ph.D. Thesis, I.A.R.I., New Delhi.

Sohal, T.S., Dubey, V.K. and Hundal, J.S. (1977) Opinion of Dairy Samachar Readers about its Utility. Indian Journal of Extension Education, 23 (1 & 2).

Somasundram, D. (1976) A Diagnostic Study of Small Farmers with Respect to New Agricultural Technology and its Effective Communication for Adoption, Ph.D. Thesis, I.A.R.I., New Delhi.

Soni, M.C. (1974) A Study of Radio Listening Habits of Farmers of Janagadh District Hearing Rural Radio Braodcast by AIR Rajkot. M.Sc. Thesis, Gujarat Agricultural University, Ahmedabad.

Stevens, K.C. (1980) The Relationship between interest and reading comprehension Dissertation Abstracts International, 40(8).

Swaminathan, M.S. (1982) The Role of Education and Research in Enhancing Rural Women's Income and Household Happiness. First J.P. Naik Memorial Lecture, New Delhi (Sept, 11th) Centre for Women's Development Studies, New Delhi.

Talukdar, R.K. and Pawar, S.S. (1981) Utility of Farm Broadcasts from AIR, Gauhati, Journal of Research Assam Agricultural University, Vol. 2(2).

Waisanen, F.B. and Durlak J.T. (1968) The Impact of Communication on Rural Development (in Costa Rica) UNESCO, Paris.

Vijayaraghavan, R. (1978) A Comparative Study of Progressive and Non-progressive Farmers' Discussion Group (Charchamandals) Organised by Farmers Training Centre, Coimbatore, Tamil Nadu, M.Sc. Thesis, IARI, New Delhi.

INDEX

Adult literacy rate in India 9
Aggarwal (1972) 38
Agricultural
  Extension 2
  labourers 7
AIR, Hyderabad (1975) 22
Ambastha (1974) 20
Annamalai (1979) 18,20
Ardashir Irani 70
Audience Research
  Cell (1976) 25
  Unit (1973,1976) 60
  Unit (1979) 19
  Unit of AIR, Delhi (1979) 17,25,26
  Unit of Madras Doordarshan Kendra (1978) 26,60
Average expectation of life at birth 6

Badrinarayana (1977) 17
Bamnauli 40
Bhandari (1972) (LIC 1972) 17,18,21,22,54
Bhani Ram (1981) 18,20, 23
Bharthal 40
Breast - feeding 83

CARE 35
Census of India, 1981 6-12,65, 81
Chahil (1972) (LIC, 1972) 35
Chandrakandan (1981) 17-19,22
Charcha - mandals 19
Chatterjee (1973) 30
Child Marriage Restraint Act of 1978, 14
Chowla (1983) 33,37,50
Co-education 78
Collection of Data and Analysis 46
Committee on status of Women in India (1974) 34,38,70,78,81
Community
  identification 3
  awareness 45
Cultivators 7

Department of Family Welfare 13
Dey (1968) 23
Dhadhal (1973) 18
Dhaliwal and Sohal (1967) 18,20
Dhillon (1968) 31,35
Divorce 81
Domestic Chores 83

Economic Issues 79
Edeani (1980) 37,48,49
Education 9,18
Educational levels 12
Equality of wages 79

Family Planning 13,82
Farm school on the AIR 17,19,20,22, 23,37
Farm Woman, 1,5
Farm Women according to their age 47
  Community awareness 49
  education level 49
  innovativeness 48
  mass media exposure 47
Farm womans
  favourite radio programmes 53
  opinion about the impact of television
  opinion on impact of radio listening 56
Feaster (1968) 44
Feeding and Childcare practices 82
Female
  Literacy 41
  mortality 5
  population 6
Films 16,33
Film Viewing Behaviour of farm Women 72

Garg (1972) (LIC,1972) 39
Govt. of India, (1974) 81
Gramin mahilaon ka karyakram 17,22,56

Hassan (1983) 36
Hindu Marriage Act (1955) 81
Hindu Women's Right to Residence and

Maintenance Act (1964) 81

I I M C (1969) (GOI, 1974) 76
Impact of
    Films 74
    Radio 55
    T.V. 60
Indian Agricultural Research Institute, New Delhi 40
Indian Constitution 78
Indian Institute of Mass Communication (1969) (GOI, 1974) 33, 34
Infant mortality level 11
Inheritance of property 77
Innovativeness 44
Intake per day of calories and Iron 13

Jalihal and Murthy (1974) 18
Jalihal and Srinivasamurthy (1974) 21
Jha (1978) 25-27,60
Joanne Leslie (1977) 35,36
John Knight (1973) 22

Kakar (1972) 39
Kamath (1973) 27
Kaur
    (1970) 24-27,60,61
    (1972) 38
    (1982) 30,32,68
Khajapeer (1978) 30
Khandekar and Mathur (1975) 29-31
Khattar (1972) (LIC,1972) 39
Kinyanjui (1972) 29-31
Kishore (1968) 22,55
Kivlin *et al* (1968) 55
Koshy and Bhagat (1978))39
Koshy and Sarin (1970) 39
Krishi-Darshan 24

Land Holding 19
Legal status 78
Lerner (1958) 1,36
Lerner (1963) 3,47,49
Life style 43
Loan for Social & Religious ceremonies 80

Magazines 16
Magazines read or heard by farm women 67
Malnutrition 13
Manjaiyon (1973) 31
Manorama Year Book, (1988) 50

Marital Status 14
Masani (1976) 18,21
Mass media
    and women 2
    exposure 3,37,45
    of communication 1
    on rural women 1
    to generate public opinion 16
    tool of 50
    use behaviour 2
Mathew (1982) 20,21
Mcnelly (1970) 3,47-49
Mehta (1972) (LIC,1972) 20-22,54,56
Micro Level Profile of farm Women 46
Ministry of Health and Family Welfare 13
Misra (1967) 23
Modern gadgets 84
Muis (1983) 27,37
Multi Media
    Approach 33,34
    Combinations 16
Murthy (1969) 24
Muslim law 81

National Council for Women Education 16
National Film Development Corporation 72
National Institute of
    Audio Visual Education (1961) (GOI, 1974) 34
    Nutrition, (1984) 13
National Readership Survey (1978) 31,68
Newspaper 16
Newspaper read or heard by farm women 67
Nigam
    (1964) (GOI,1974) 34
    (1987) (Dte Extn 1987) 23

Occupational pattern of female workers 8

Patel (1976) 19
Pillai and his associates(1974) 26
Pilot study 42
Pochanpur 40
Polygamy 80
Preferential film viewing pattern 75
Pregnant women 14
Press in India (1984) 66
Print Media 65
Profile of Radio Listeners 17
Programme preference 21,53,59
Puri (1972) 17,20-22,35,54-56

INTRODUCTION

Radio 16,50
Radio as a source of Information 20
Radio Listening Behaviour 20
Rajamani (1981) 22,37
Ramson (1977) 36
Reason's for
  owning radio sets 52
  using print media 68
Reddy (1971) 25
Registrar of Newspapers 30
Relevance of films 73
Religious films 75
Research strategy 40
Rights to Property 79
Rogers (1969) 1,3,36,44,47-49
Romantic films 75

Sadamate (1975) 24,25,27,61,
Saini (1985) 68
Sakya (1973) 17,18,20,21
Sandhu (1970) 17-21
Snoria and Singh (1976) 31,68
Sarkar (1982) ,65,68
Saroj Malik (1987) (Dte Extn 1987) 27
Satellite Instructional Television
  Experiment 27
Saving money for old age 79
Schneider (1970) 31
Schwartz (1981) 31,37,43,49,66
Sekhon (1972) 24,25,27,60
Seventh Five Year Plan (1990) 57,72
Shahi (1977) 60,61
Shankariah (1969) 20
Sharda Act 14
Shivpuri (1972) 29,31
Singh (1972) 17,21,24,25,27,38,60,61
Singh and Sandhu (1971) 18
Sinha (1974) 24,27,60,61
Sita and Krishnan (1975) 26
Social
  Issues 44,77,78
  participation 18
Socio-economic and political status of women 77

Socio-family drama 75,76
Sohal *et al* (1977) 31
Sohi (1973) 24,25
Somasundram (1976) 20
Soni (1974) 19
State and Union territories in order of female
  literacy, (1981) 9
Status of Health 77
Stevens (1980) 30
Studies on Newspapers and Magazines 29
Studies Related to Television 23
Sundarajan *el at* (1978) 22
Supe (1971) 31
Supplementary feeding 83
Survey (1957) (GOI,1974) 34
Survey conducted in Bombay city (1957)
  (GOI,1974) 76
Survey of Infant & child Mortality, (1979) 11

Talukdar and Pawar (1981) 18,20
Tamil Nadu Agricultural University Coimbatore (1978) 17
Television 16,57
Trends in female literacy in India 11
TV
  Centre Srinagar (1975) 26
  Dilemma 57
  Ownership 58
  Programme preferences expressed by
    farm women 61

Vijayaraghavan (1978) 17,19,20,22
Voting 84

Waisanen and Durlak (1968) 22,55
Widow remarriage, 80
Work Force and Occupation 7
Work participation rates 7
World Bank, (1982) 6

Yadav (1983) 66
Yuv-vani 21,22